*Science Assessment Leadership T*
*Jan 18-19, 2003 at*
*Eastern Washington University*

Systems

DESIGN

# Fender Bender Physics

By Roy Q. Beven and Robert A. Raudebaugh

*Heather, Thanks for hosting ed. reform*

*Royd Beven*   *Robert B Raudebaugh*

THE WORLD'S A CLICK AWAY

Featuring SciLinks—a new way of connecting text and the Internet.
Up-to-the-minute online content, classroom ideas, and other materials
are just a click away.

**NSTApress**
NATIONAL SCIENCE TEACHERS ASSOCIATION
Arlington, Virginia

Published with support from The National Highway Traffic Safety Administration

**NSTA press**
NATIONAL SCIENCE TEACHERS ASSOCIATION

**Press Staff**
Judy Cusick, Associate Editor
Carol Duval, Associate Editor

**Art and Design**
Linda Olliver, Director

**NSTA Web**
Tim Weber, Webmaster

**Periodicals Publishing**
Shelley Carey, Director

**Printing and Production**
Catherine Lorrain-Hale, Director

**Publications Operations**
Erin Miller, Manager

**sciLINKS**
Tyson Brown, Manager

**National Science Teachers Association**
Gerald F. Wheeler, Executive Director
David Beacom, Publisher

NSTA Press, NSTA Journals, and the NSTA Web site deliver high-quality resources for science educators.

*Fender Bender Physics*
Library of Congress Catalog Card Number 2001093751
NSTA Stock Number: PB144X2
ISBN 0-87355-190-7
Printed in the United States of America by Victor Graphics
Printed on recycled paper

NSTA Press
1840 Wilson Boulevard
Arlington, VA  22201-3000
http://www.nsta.org/

National Science Teachers Association

People Saving People
www.nhtsa.dot.gov

# Acknowledgments

I thank my wife, Michele, and children, Lauren and Drew, for their support throughout the long process of creating this integrated curriculum. I gained a tremendous amount of middle school wisdom while writing and field-testing all of *Fender Bender Physics* at Kulshan Middle School and the Spaceframe Unit at Bellingham High School in Bellingham, Washington. For what they taught me, I thank Sherrie Brown, Kulshan's principal, and my fellow teachers of the 1999-2000 7th grade Sentinel Team, Marv Fast, Marion Hiller, and Peggy Zender.
—*Roy Q. Beven*

I would like to acknowledge all of my students in the Technology Education classes at Western Washington University who patiently endured my experimentation and lectures on the benefits of integrating math and science concepts into design and technology activities. I would also like to acknowledge my wife, Linda, without whose patience I may never have finished this manuscript.
—*Robert A. Raudebaugh*

*Fender Bender Physics* was reviewed by Linda Kralina, science coordinator for Rockwood School District in Eureka, Missouri; Jeffrey Leaf, technology teacher at Thomas Jefferson High School for Science and Technology, Fairfax County, Virginia; Ron Morse, science teacher at Minoa High School, East Syracuse, New York; Barbara Starkey, science teacher at Clair E. Gale Junior High School in Idaho Falls, Idaho; Roy Unruh, professor of physics and science education at the University of Northern Iowa; and Dean Zollman, professor of physics at Kansas State University in Manhattan, Kansas. Activities in this book were field-tested by Dot Dickinson, physical science teacher at Episcopal Middle School in Baton Rouge, Louisiana.

Cheryl Neverman of the National Highway Traffic Safety Administration provided guidance and encouragement throughout the development of *Fender Bender Physics*.

*Fender Bender Physics* is produced by NSTA Press: Shirley Watt Ireton, director; Judy Cusick, associate editor; Carol Duval, associate editor. Beth Daniels was project editor for *Fender Bender Physics*. Juliana Texley, Barbara Seeber, and Georgia Martin provided additional editing assistance. Also at NSTA, Anne Early, Sara Gebhardt, Jessica Green, Erin Miller, and Catherine Lorrain-Hale provided invaluable support throughout the production of this book. Viviane Moritz of Graves Fowler Associates designed the book and cover from a cover illustration by Jeffrey Pelo. The inside illustrations for this book were created by Viviane Moritz of Graves Fowler Associates from originals by Roy Beven and Robert Raudebaugh.

# Table of Contents

Acknowledgments . . . . . . . . . . . . . . . . . . . . . . . . . . . . . . . . . iii

Correlations with the *National Science Education Standards* . . . . . . vi

SciLinks . . . . . . . . . . . . . . . . . . . . . . . . . . . . . . . . . . . . . vii

## Introduction

Overview . . . . . . . . . . . . . . . . . . . . . . . . . . . . . . . . . . . . . . 1

Curriculum Design . . . . . . . . . . . . . . . . . . . . . . . . . . . . . . . . 2

Meeting Standards in Science and Technology . . . . . . . . . . . . . . . . 3

Implementing the Curriculum . . . . . . . . . . . . . . . . . . . . . . . . . . 6

Communicating Science and Technology . . . . . . . . . . . . . . . . . . . . 8

Spreading the Word . . . . . . . . . . . . . . . . . . . . . . . . . . . . . . . . 9

## Unit 1—The Mousetrap Car

**Teacher's Guide to Unit 1** . . . . . . . . . . . . . . . . . . . . . . . . . . . 11

**Introduction:** The Mousetrap Car Challenge
(Student Handout) . . . . . . . . . . . . . . . . . . . . . . . . . . . . . . . . . 19

**Activity 1:** Writing Design Briefs . . . . . . . . . . . . . . . . . . . . . . . 21

**Activity 2:** Measuring Force . . . . . . . . . . . . . . . . . . . . . . . . . . 26

**Activity 3:** Writing a Design Process Paper . . . . . . . . . . . . . . . . . 31

**Activity 4:** Testing Wheels and Axles . . . . . . . . . . . . . . . . . . . . 34

**Activity 5:** Measuring Lever-Arm Effects . . . . . . . . . . . . . . . . . . 39

**Activity 6:** Developing a Design . . . . . . . . . . . . . . . . . . . . . . . . 44

**Activity 7:** Constructing a Prototype . . . . . . . . . . . . . . . . . . . . 47

**Activity 8:** Measuring Motion . . . . . . . . . . . . . . . . . . . . . . . . . 50

## Unit 2—The $CO_2$ Car

**Teacher's Guide to Unit 2** . . . . . . . . . . . . . . . . . . . . . . . . . . . 57

**Introduction:** The $CO_2$ Car Challenge (Student Handout) . . . . . . . . 64

**Activity 1:** Design Requirements and Constraints . . . . . . . . . . . . . 66

**Activity 2:** Mass and Motion . . . . . . . . . . . . . . . . . . . . . . . . . 70

**Activity 3:** Force Diagrams . . . . . . . . . . . . . . . . . . . . . . . . . . . . . . . . 76

**Activity 4:** Frictional Force . . . . . . . . . . . . . . . . . . . . . . . . . . . . . . . . . 82

**Activity 5:** Writing Design Briefs . . . . . . . . . . . . . . . . . . . . . . . . . . . 90

**Activity 6:** Prototype Development . . . . . . . . . . . . . . . . . . . . . . . . . . 93

**Activity 7:** Performance Testing . . . . . . . . . . . . . . . . . . . . . . . . . . . . 98

**Activity 8:** Time and Velocity . . . . . . . . . . . . . . . . . . . . . . . . . . . . . 100

## Unit 3—The Space-Frame Vehicle

**Teacher's Guide to Unit 3** . . . . . . . . . . . . . . . . . . . . . . . . . . . . . . . 105

**Introduction:** The Space-Frame Vehicle Challenge
(Student Handout) . . . . . . . . . . . . . . . . . . . . . . . . . . . . . . . . . . . . . . . 111

**Activity 1:** Passenger Safety . . . . . . . . . . . . . . . . . . . . . . . . . . . . . . 113

**Activity 2:** Force-Path Diagrams I . . . . . . . . . . . . . . . . . . . . . . . . . 117

**Activity 3:** Force-Path Diagrams II . . . . . . . . . . . . . . . . . . . . . . . . . 124

**Activity 4:** Testing Material Properties . . . . . . . . . . . . . . . . . . . . . . 131

**Activity 5:** Design Documentation . . . . . . . . . . . . . . . . . . . . . . . . . 136

**Activity 6:** Work and Energy . . . . . . . . . . . . . . . . . . . . . . . . . . . . . . 139

**Activity 7:** Destructive Testing and Analysis . . . . . . . . . . . . . . . . . 145

**Activity 8:** Energy Transfers . . . . . . . . . . . . . . . . . . . . . . . . . . . . . . 148

## Supplemental Materials

**Reading 1:** Brainstorming . . . . . . . . . . . . . . . . . . . . . . . . . . . . . . . . 153

**Reading 2:** The Design Process . . . . . . . . . . . . . . . . . . . . . . . . . . . . 156

**Reading 3:** Mind Mapping . . . . . . . . . . . . . . . . . . . . . . . . . . . . . . . 162

**Reading 4:** Motion and Force . . . . . . . . . . . . . . . . . . . . . . . . . . . . . 164

**Reading 5:** Student Safety Guide . . . . . . . . . . . . . . . . . . . . . . . . . . 171

**Reading 6:** Work as Mechanical Energy . . . . . . . . . . . . . . . . . . . . . 174

**Reading 7:** The Manufacturing Process . . . . . . . . . . . . . . . . . . . . . 179

**Sketching Guide** . . . . . . . . . . . . . . . . . . . . . . . . . . . . . . . . . . . . . . . 181

# Correlations with the *National Science Education Standards* for Grades 5–8

| CONTENT STANDARD | TOPIC | MOUSETRAP CAR ACT 2 | ACT 3 | ACT 4/7 | ACT 5 | ACT 8 | CO₂ CAR ACT 1 | ACT 2 | ACT 3 | ACT 4 | ACT 6/7 | ACT 8 | SPACEFRAME VEHICLE ACT 1 | ACT 2 | ACT 3 | ACT 4 | ACT 6 | ACT 7 | ACT 8 |
|---|---|---|---|---|---|---|---|---|---|---|---|---|---|---|---|---|---|---|---|
| (A) Science as Inquiry | Develop descriptions, explanations, predictions, and models using evidence | | | • | | | | | | | • | | | | | • | | • | |
| | Communicate scientific procedures and explanations | | • | | | • | | | | | • | • | | | | | | • | • |
| | Understandings about scientific inquiry | | • | | • | • | • | | | | • | • | • | | | • | | • | • |
| (B) Physical Science | Motion and Forces | • | | | • | | | • | • | • | | | | • | • | | | | |
| | Transfer of Energy | | | | | | | | | | | | | • | • | | • | | • |
| (E) Science and Technology | Abilities of technological design | | • | • | | | • | | | | • | • | • | | | • | | • | |
| | Understanding about science and technology | | | | | | • | | | | • | | • | | | | | | |
| (F) Science in Personal And Social Perspectives | Science and technology in society | | | | | | | | | | • | | • | | | | | • | • |
| (G) History and Nature of Science | Nature of science | | • | | | | | | | | | | • | | | | | | |

National Research Council. *National Science Education Standards*. Washington, DC: National Academy Press, 1996.

*Fender Bender Physics* brings you SciLinks, a new project that blends the two main delivery systems for curriculum—books and telecommunications—into a dynamic new educational tool for all children, their parents, and their teachers. This effort, called SciLinks, links specific science content with instructionally rich Internet resources. SciLinks represents an enormous opportunity to create new pathways for learners, new opportunities for professional growth among teachers, and new modes of engagement for parents.

In this SciLinked text, you will find an icon near several of the concepts you are studying. Under it, you will find the SciLinks URL (*www.scilinks.org*) and a code. Go to the SciLinks Web site, sign in, type the code from your text, and you will receive a list of URLs that are selected by science educators. Sites are chosen for accurate and age-appropriate content and good pedagogy. The underlying database changes constantly, as dead or revised sites are eliminated or replaced with better selections. The ink may dry on the page, but the science it describes will always be fresh. SciLinks also ensures that the online content teachers count on remains available for the life of this text. The SciLinks search team regularly reviews the materials to which this text points—revising the URLs as needed or replacing Web pages that have disappeared with new pages. When you send your students to SciLinks to use a code from this text, you can always count on good content being available.

The selection process involves four review stages:

1. First, a cadre of undergraduate science education majors searches the World Wide Web for interesting science resources. The undergraduates submit about 500 sites a week for consideration.

2. Next, packets of these Web pages are organized and sent to teacher-Webwatchers with expertise in given fields and grade levels. The teacher-Webwatchers can also submit Web pages that they have found on their own. The teachers pick the jewels from this selection and correlate them to the *National Science Education Standards*. These pages are submitted to the SciLinks database.

3. Then scientists review these correlated sites for accuracy.

4. Finally, NSTA staff approve the Web pages and edit the information provided for accuracy and consistent style.

SciLinks is a free service for textbook and supplemental resource users, but obviously someone must pay for it. Participating publishers pay a fee to the National Science Teachers Association for each book that contains SciLinks. The program is also supported by a grant from the National Aeronautics and Space Administration (NASA).

# Introduction
## Overview

*Fender Bender Physics* is an integrated science and technology curriculum built on the theme of transportation safety. The book contains three units that can be used together, individually, or as enrichment in a standard middle school physical science program. All three units are based on the *National Science Education Standards* for appropriate grade-level concepts, processes, and skills.

In *Fender Bender Physics* students build three progressively more complex vehicles. Students use scientific inquiry to gather information critical to vehicle design. Like teams of professional engineers, student teams prepare design portfolios and present their projects to the class. The eight instructional activities in each unit help students develop technological skills necessary to build, test, and evaluate their prototypes. Students acquire core concepts, develop an understanding of the design process, and increase their technical skills through the material presented in each unit.

Throughout *Fender Bender Physics* connections are made to vehicle and driving safety. In "Unit 1—The Mousetrap Car," students examine inertia and the potential for a body (car or occupant) to remain in motion once it accelerates. In "Unit 2—The $CO_2$ Car," students compare the capability of vehicles carrying a variety of loads to remain on a track (road) and examine the importance of wheel alignment. Students diagram vehicles' centers of gravity, in order to compare the stability of sport utility vehicles with traditional sedans. In "Unit 3—The Space-Frame Vehicle," students look at how the design of a car chassis protects its occupants. In each unit extensions appropriate for independent research are provided.

**Curriculum Design and Use**

The curriculum of each unit is based upon a seven-phase design process (see figure, next page).

## Curriculum Design

Each unit is based on a seven-phase design process.

### SEVEN-PHASE DESIGN PROCESS

| DEFINE THE PROBLEM |
|---|

| SCIENTIFICALLY GATHER INFORMATION | EXPLORE IDEAS |
|---|---|

| DEVELOP THE DESIGN |
|---|

| CONSTRUCT PROTOTYPE | SCIENTIFICALLY TEST AND EVALUATE |
|---|---|

| REDESIGN |
|---|

**Team Project Presentation**

Within each unit clear connections are drawn between the eight instructional activities and the seven phases of the design process. Units begin with an overview that includes a discussion of the design challenge, the requirements for design, a list of the instructional activities, and unit assessment criteria.

The material is especially valuable for schools struggling to raise achievement levels in physical science without eliminating valuable technology education experiences. Materials provide real-world applications that are especially interesting to middle school students.

Each unit and instructional activity has a teacher's guide along with student pages designed for reproduction as in-class materials. A unit teacher's guide introduces each unit, discusses conceptual development, and provides suggestions for implementation, time management, and assessment. *Fender Bender Physics* also contains a section of supplemental reading for middle school students. These materials include the following:

- a sketching guide

- background on the physics of motion, force, work, and on the manufacturing process

- techniques, such as brainstorming and mind mapping, that can be employed in the process of solving problems

- detailed descriptions of various stages of the design process

- a student safety manual that outlines overall safety procedures and provides illustrations of the various tools and machines used in the curriculum, including specific safety considerations pertinent to each

## Meeting Standards in Science and Technology

*National Science Education Standards* encourages "activities that are meant to meet a human need, solve a human problem, or develop a product" [National Research Council (NRC), 161]. Technology education provides middle level students with activities that allow them to develop "real technological products, systems, and environments" and to create models [International Technology Education Association (ITEA), 38]. Many schools have elective technology programs. But this curriculum integrates science and technology in a way that meets the standards of both areas.

In the construction of prototype vehicles, *Fender Bender Physics* meets the following standards (NRC, 165):

- Identify appropriate problems for technological design.

- Design a solution or product.

- Implement a proposed design.

- Evaluate completed technological designs or products.

The program builds upon the following science content standards appropriate for the middle school student (NRC, 154-57):

- The motion of an object can be described by its position, direction of motion, and speed. The motion can be measured and represented on a graph.

- An object that is not being subjected to a force will continue to move at a constant speed and in a straight line.

- If more than one force acts on an object along a straight line, then the forces will reinforce or cancel one another, depending on their direction and magnitude. Unbalanced forces will cause changes in the speed or direction of an object's motion.

- Energy is a property of many substances and is associated with heat, light, electricity, mechanical motion, sound, nuclei, and the nature of a chemical. Energy is transferred in many ways.

The program encourages process skills as well. "Different systems of measurement are used for different purposes. Scientists usually use the metric system. An important part of measurement is knowing when to use what system" (NRC, 118). In *Fender Bender Physics*, the metric system is employed.

The International Technology Education Association has established standards parallel to those established by the National Research Council. The recommended standards for middle school students (ITEA, 95) in grades six, seven, and eight include the following:

- Design is a creative planning process that leads to useful products and systems.

- There is no perfect design.

- Requirements for a design are made up of criteria and constraints.

Other ITEA recommendations for middle school students include the following:

To comprehend engineering design (ITEA, 103):

- Design involves a set of steps, which can be performed in different sequences and repeated as needed.

- Brainstorming is a group problem-solving design process in which each person in the group presents his or her ideas in an open forum.

- Modeling, testing, evaluating, and modifying are used to transform ideas into practical solutions.

To comprehend other problem-solving approaches (ITEA, 110):

- Troubleshooting is a problem-solving method used to identify the cause of malfunction in a technological system.

- Invention is a process of turning ideas and imagination into devices and systems.

- Innovation is the process of modifying an existing product or system to improve it.

- Some technological problems are best solved through experimentation.

To use and maintain technological products and systems (ITEA, 130):

- Use information provided in manuals, protocols, or by experienced people to see and understand how things work.

- Use tools, materials, and machines safely to diagnose, adjust, and repair systems.

- Use computers and calculators in various applications.

- Operate and maintain systems in order to achieve a given purpose.

To select, use, and understand energy and power technologies (ITEA, 162):

- Energy is the capacity to do work.

- Energy can be used to do work using many processes.

- Much of the energy used in our environment is not used efficiently.

To select, use, and understand information and communication technologies (ITEA, 172):

- The use of symbols, measurements, and drawings promotes clear communication by providing a common language to express ideas.

To select, use, and understand transportation technologies (ITEA, 171):

- Transportation vehicles are made up of subsystems, such as structural, propulsion, suspension, guidance, control, and support, which must function together for a system to work effectively.

All the instructional activities in *Fender Bender Physics* "engage students in identifying and shaping an understanding of the question under inquiry" (NRC, 144). Building their projects, students "know what the question is asking, what background knowledge is being used to frame the question, and what they will have to do to answer the question" (NRC, 144). Accordingly, the following middle level standards (NRC, 145-148) are developed in the activities presented in this book:

- Identify questions that can be answered through scientific investigations.

- Design and conduct a scientific investigation.

- Use appropriate tools and techniques to gather, analyze, and interpret data.

- Develop descriptions, explanations, and models using evidence.

- Think critically and logically to make connections between evidence and plausible explanations for that evidence.

- Recognize and analyze alternative explanations and predictions.

- Communicate scientific procedures and explanations.

- Use mathematics in all aspects of scientific inquiry.

## Implementing the Curriculum

To make the most of *Fender Bender Physics*, teachers should have a good understanding of both science and technology. A partnership with the school's technology department provides excellent connections. The program works well in an inclusive classroom environment. We have included specific suggestions for improving rapport between teachers and students, for making students feel that their contributions are wanted and needed, for communicating a sense of excitement about the subject matter and the activities, and for bringing out the best in every student in terms of effort and participation.

Engineering is still perceived by some as a field for white males. Similarly, today's technology teachers may often be white males who have not had diverse science experiences. Instructional activities are specifically designed to strengthen a range of skills and build confidence in students who may not have considered technology careers. The activities are also scientifically rigorous, a first step toward dispelling the invisible line between college preparatory curricula and technology curricula.

In *Fender Bender Physics* students work in cooperative groups called Vehicle Research Teams (VRT) to simulate the workplace. Teams define the problems, do research, and run experiments to gather information for their designs. Next, teams explore ideas through sketches, tests, experiments, and group discussions. Finally, they decide on a design and get it approved by the teacher. VRTs construct working models of their design prototypes and then test and evaluate their prototypes, redesigning them as needed. This system not only saves time but also reflects modern engineering practices.

Middle school students need to develop the interpersonal skills that enable them to work cooperatively. The instructional activities of this curriculum suggest ways that students can collaborate with their peers in working through the various tasks involved, then write up the results individually. Suggestions for team roles and responsibilities are given.

The teacher must decide how many groups or teams a facility can support. This will depend upon space, furniture, and equipment. Smaller teams allow more opportunities for full participation; larger groups require fewer materials and are easier to supervise. Because space is an important factor in safety, some teachers choose to use a larger space (such as a technology laboratory or cafeteria) for the final stages of construction.

Many teachers form heterogeneous cooperative teams of three or four students who have been through a series of team-building activities together. It is important to avoid homogeneous groups that are all highly skilled, all inexperienced, or all poorly motivated. In larger teams teachers should see that jobs are fairly distributed and that girls have equal opportunities for construction tasks.

## Communicating Science and Technology

Technical writing is a significant component of *Fender Bender Physics*. Each instructional activity requires students to communicate ideas, explanations, solutions, and connections. In addition, students are asked to keep classroom notes in a notebook. The organization of the notebook is a significant student responsibility. Many activities end with a reminder to students to record the definitions of scientific and technological concepts and processes presented in class in their notebooks.

Finally, every student is asked to assemble a design portfolio to fulfill the following purposes:

- Preserve important thoughts, feelings, and experiences from a unit.

- Show growth as a writer, scientist, engineer, and/or technician during a unit.

- Display important pieces of work for a unit.

- Showcase the whole range of their writing, mathematical, and sketching skills from a unit.

- Reflect upon the design process and ideas developed in a unit.

The design portfolio for each unit is a repository of students' experiences. A design portfolio includes some material that may have originally appeared in the notebook, rewritten in a more formal style intended for others to read. A design portfolio also includes an expository paper composed of several paragraphs describing how the instructional activities of the unit affected the design process. Portfolios allow students to become comfortable with the whole notion of documenting a process.

Sketching and technical drawing are important to every unit. Most curricula do not include a technical drawing component, yet such skills are essential in many careers. Through these activities students develop the ability to use sketching as a tool to get ideas onto paper. Drawing techniques allow students to explore a wide range of ideas by using sketching as a form of idea shorthand. Drawing also supports diverse learning styles and encourages students who may not have been successful in science in the past.

## Spreading the Word

This curriculum deals with real transportation issues that challenge modern engineers and challenge our scientific literacy. From skates and scooters to interstellar rockets, transportation makes news. Your entire school will be interested in these connections. Each product of *Fender Bender Physics* has close connections to other inventions. Mousetrap cars work much like medieval catapults. Compressed air is used in the $CO_2$ car; compressed air also propels the tube in a drive-up bank teller machine. Space-frame structures in cars save lives in head-on collisions. The history of automobile safety is rich in the contributions of various cultures. These connections make excellent displays, newsletters, and community press releases. You can enhance the value of the culminating activities in each unit by using videotape or digital movies of the runs so that students can analyze them at a later date. Digital movies can be incorporated into student Power Point presentations or webpages on automobile safety.

The program is especially valuable when teachers share their expertise with one another. Traditional science teachers can learn from professional development in technology education to gain an understanding of design and engineering. Similarly, technology teachers can learn from the processes of scientific inquiry and physics concepts. Partnerships enrich curricular connections and provide continuing growth.

The following sources of information can be useful:

- National Science Teachers Association (NSTA): *www.nsta.org*

- National Highway Traffic Safety Administration (NHTSA): *www.nhtsa.dot.gov*

- National Science Foundation (NSF): *www.nsf.gov*

- American Association of Physics Teachers (AAPT): *www.aapt.org*

- International Technology Education Association (ITEA): *www.iteawww.org*

## References

International Technology Education Association. *Standards for Technological Literacy*. Reston, VA: ITEA, 2000.

National Research Council. *National Science Education Standards*. Washington, DC: National Academy Press, 1996.

# Teacher's Guide to Unit 1:
## The Mousetrap Car

*The integration of science and technology can provide exciting learning opportunities for the middle school student. In this unit students design and build a vehicle that employs a mousetrap as a source of energy. Students participate in basic research, design, construction, and evaluation in a process of continuous inquiry, supported by record keeping and effective communication.*

## Overview

Students begin by investigating the force of a mousetrap's lever arm. They measure the relationship between the distance a mousetrap is opened and the force it exerts as well as the effect of the snapper-arm length on that force. Then students apply their findings to the design of a mousetrap car. To evaluate, students measure, graph, and describe the motions of their designs. Students read background information on force and motion and communicate their results to others.

Like a professional scientific investigation, this project relies upon the accumulation of accurate data. Students create their own database of definitions, concepts, and experimental results. Students create design requirements and specifications and learn how to incorporate sufficient detail in order to get a design approved. Students also develop sketching skills as a tool for designing and communicating ideas.

Finally, students gain the satisfaction of seeing their designs take shape from a prototype. Prototyping is a standard engineering process that involves testing a variety of materials to determine which ones best meet the requirements of a project. Working in teams, students evaluate data by comparing car designs and communicate their progress through reports and expository writing.

This unit on the mousetrap car shows students the dynamic relationship between scientific investigation and technology. Students use experimentation, testing, and evaluation at each step of the process. They learn that their progress must be measurable and

## Materials

- Copy of student handout, "The Mousetrap Car Challenge," for each student

- Copies of each of the eight activities, plus *Brainstorming, Mind Mapping, The Design Process,* and *Sketching Guide* for each student

- One mousetrap per team

- One 10-N spring scale per team

- One 2.5-N spring scale per team

- One set of small wooden blocks (approximately 1 cm$^2$, 1.5 cm$^2$, 2 cm$^2$, and 2.5 cm$^2$) per team

- One meterstick per team

- One stopwatch per team

- One wheel-block assembly per team (2.5-cm x 10-cm wood block)

- One 61-cm x 244-cm piece of plywood

- Test block made of pine (15 cm in length with 2.5-cm x 10-cm cross section, equipped with holes and bushings) per team

*continued on next page*

## Materials *continued*

- Cardboard to stop the test block

- Drinking straws

- Four axle rods (wood, plastic, and metal—all about 5 mm in diameter and 12 cm in length) per team

- Two sections of brass tubing (10 cm in length and with an inside diameter of about 6 mm) per team

- Enough materials for each team's wheels, axles, and bushings (CDs, old records, lids, and anything else that is round)

- Large ball of string

- Chassis materials (wood scraps, tongue depressors, strips of plastic, metal rods)

- Hand tools (NO power tools in the middle school)

- Material for lever-arm extension (optional)

- Eye protection

must be communicated in operational terms. When questions arise, controlled investigations provide answers. A design portfolio is the authentic assessment of the inquiry process.

The culminating activity of the unit is a Great Race. If students race their vehicles, emphasize pride over competition. The unit should be fun—ending with friendly, cooperative team presentations of design portfolios and completed mousetrap vehicles.

## Conceptual Development

To launch the unit, introduce the challenge of designing and building a mousetrap car by distributing the student handout, "The Mousetrap Car Challenge." The handout gives students an overview of expectations and a table for recording points earned. Tell students to make the handout the first page of their notebooks.

Don't assume your students have ever seen a mousetrap or a mousetrap car. Begin by building a mousetrap car yourself. Construct the simplest version of the vehicle and share it with the class. Let them see how it works. Some groups may build cars very similar to this basic version, but most will add their own innovations. In "Activity 1: Writing Design Briefs," you will show them how a mousetrap car operates—and how strong (and potentially dangerous) the spring of a mousetrap is. Let the arm close on a marshmallow to demonstrate the potential danger to a hand. Tell students that horseplay will not be tolerated.

Students begin their study by defining the goals and design requirements of building a mousetrap car. As apprentice engineers, they learn that the design process requires many sequential steps and careful analysis. They practice group brainstorming and begin systematic record keeping.

In "Activity 2: Measuring Force," you will have students measure the force exerted by a mousetrap. The manipulated variable is the distance the trap is opened. Students will discover that the force is directly proportional to the stress within the normal range of the trap. This is Hooke's law, which applies to all elastic materials, including the spring steel of the trap. This relationship can be incorporated into their mousetrap-car design. In later activities students investigate the effects of lengthening the lever arm on the trap and make decisions about the value of this modification in comparison to the added mass. These first controlled investigations should be conducted slowly and deliberately. Many middle schoolers are just beginning to develop the logical skills that they need to understand variables, controls, and proportion. Emphasize the importance of accuracy and careful measurement, and reinforce the connections between the experiments and the design process.

In Activity 3 students begin the design process using the data they have collected to develop sketches and specifications for their own design. In later activities they test materials for axles and bushings and develop consensus on design issues. They are encouraged to invent creative solutions to the challenge defined by their goals and product requirements. They document their progress in a notebook and ultimately in a design portfolio. The diagram below shows the instructional activities and how they link with the design process.

## Mousetrap Design Challenge
### Design Process With Associated Instructional Activities

```
                    ┌──────────────────────────────────┐
                    │        DEFINE THE PROBLEM        │
                    │  Activity 1: Writing Design Briefs │
                    └──────────────────────────────────┘

┌─────────────────────────────────────┐   ┌──────────────────────────────────┐
│  SCIENTIFICALLY GATHER INFORMATION  │   │          EXPLORE IDEAS           │
│    Activity 2: Measuring Force      │   │ Activity 4: Testing Wheels and Axles │
│ Activity 5: Measuring Lever-Arm Effects │ └──────────────────────────────────┘
└─────────────────────────────────────┘

                    ┌──────────────────────────────────┐
                    │        DEVELOP THE DESIGN        │
                    │ Activity 3: Writing a Design Process Paper │
                    │   Activity 6: Developing a Design │
                    └──────────────────────────────────┘

┌─────────────────────────────────────┐   ┌──────────────────────────────────┐
│       CONSTRUCT PROTOTYPE           │   │  SCIENTIFICALLY TEST AND EVALUATE │
│ Activity 7: Constructing a Prototype │   │  Activity 8: Measuring Motion     │
└─────────────────────────────────────┘   └──────────────────────────────────┘

                    ┌──────────────────────────────────┐
                    │            REDESIGN              │
                    │  Activity 3: Writing a Design    │
                    │        Process Paper             │
                    └──────────────────────────────────┘
```

**Team Project Presentation**

## Tips for Making the Unit Work

*Arrange the classroom for movement and active learning.* If you have access to a technology lab, such as an old shop, you may wish to do the final construction in that larger space. Assign teams, and emphasize the roles of each member.

*Collect a variety of materials to spark creativity.* Many teachers publish wish lists in their school newsletters in advance of teaching the unit. Invest in plastic bins, or ask the school office to save paper boxes. Since safety depends upon organization, emphasize to students that all excess material must be discarded properly or returned to the appropriate bin for someone else to use. If your high school has a wood shop, you may be able to get students to create the wood blocks you need there.

*Divide your students into teams carefully.* Do not put advanced students or disruptive students together. Post a list of Vehicle Research Teams (VRTs) in advance, and emphasize that each team will be graded on how well they work together. Emphasize that responsibilities must be rotated so that everyone has an opportunity for each role.

*Communicate high expectations.* Tell students that you know they are capable of earning full credit on each activity. Discretely return incomplete work for improvement. Be very specific about the requirements of the portfolio. Review graphing and paragraphing skills, so that each student can achieve success.

*Give praise and encouragement.* Communicate affection with humor, self-disclosure, eye contact, and smiles. Because you care, students will work more carefully. To model real scientific team-work, emphasize sharing rather than competing.

*Give students responsibility.* Allow students to demonstrate pride in their work by having sample mousetrap cars on display. Invite other adults into the room, and have them sit with a team while the students work. At the end of a class period, have students clean up the room and assume other housekeeping tasks. Rotate these jobs among the members of the class on a regular basis throughout the course of this unit.

*Treat all students equally and fairly.* Some students will be amazed that their cars actually roll at all, and others may be disappointed that their cars aren't doing as well as they hoped. Remind them that it is the process, not the product, that is important.

## Physics Background

Most physics and engineering studies use the same conventions— definitions, terminology, symbols, etc. They help ensure that scientists around the world and over time speak the same language. Your students will have to learn this language too. Do not begin by giving out all the definitions as notes; teach definitions opportunistically, as needed. Encourage students to create a glossary section in their notebooks for definitions and symbols used in their force and motion investigations, such as those listed below, rather than scattering definitions and symbols in many locations.

**Distance (d):** the measure of the space between two objects or events, measured in meters (m)

**Time (t):** the measure of the interval between two events, measured in seconds (s)

**Velocity (v):** the rate at which distance increases over time plus the direction, measured in meters per second (m/s). **Formula: $v = d/t$**. (Remember that velocity must include a direction. All of these tests involve an increase in distance or a positive velocity.)

**Force (F):** a push or a pull, anything that causes an object to speed up, slow down, turn, or change shape, measured in newtons (N)

**Friction ($F_f$):** the force that opposes the movement of one substance over or through another, measured in newtons (N)

**Lever Arm:** the distance from the place a force acts to the place it pivots, measured in meters (m)

**Machine:** any device that can reduce a force (by increasing the distance over which it acts) or decrease the distance (by increasing the force). **$(F)(d) = (f)(D)$**. The mousetrap is a machine with a force that is smaller when the snapper (lever) arm is close to the spring and larger as the snapper arm is extended.

**Product Requirement:** the parameters specified (by the consumer or the challenge) for the product

**Product Specification:** the specific characteristics required to make an individual design work

## Time Management

A suggested schedule for covering all the activities of this unit over the course of 20 school days (four five-day weeks) is given in the table below. It is based on class periods of 45 minutes.

### Time Management for Mousetrap Car Unit

| DAY NUMBER | ACTIVITY |
| --- | --- |
| 1 | Mousetrap Car Challenge Unit Overview |
| 2 | Activity 1: Writing Design Briefs |
| 3 | Activity 2: Measuring Force |
| 4 | Finish Activity 2: Measuring Force; Presentation: Force and Motion |
| 5 | Activity 3: Writing a Design Process Paper |
| 6 | Activity 4: Testing Wheels and Axles |
| 7 | Portfolio Presentation; Activity 5: Measuring Lever-Arm Effects |
| 8 | Finish Activity 5: Measuring Lever-Arm Effects; Presentation: Force Transformers |
| 9 | Activity 6: Sketching—Developing a Design |
| 10 | Finish Activity 6: Sketching—Developing a Design; Approve Designs |
| 11 | Activity 7: Constructing a Prototype; Portfolio Presentation |
| 12 | Design and Build Mousetrap Car |
| 13 | Design and Build Mousetrap Car |
| 14 | Design and Build Mousetrap Car; Portfolio Presentation |
| 15 | Activity 8: Measuring Motion |
| 16 | Finish Activity 8: Measuring Motion; Revisit Activity 3: Writing a Design Process Paper |
| 17 | Mousetrap Car Great Race |
| 18 | Portfolio Presentation |
| 19 | Presentations; Portfolios Due |
| 20 | Presentations; Portfolios Due |

## Unit Assessment Criteria

The most important reasons for assessment are to judge student understanding and to evaluate the effectiveness of instruction. Grades are secondary; hence, assessments that are embedded in a unit are far more valuable than tests coming at the end. Embedded assessments also are more likely to reflect the kind of knowledge that can be applied to real-world situations.

The process of designing and building a mousetrap car has few constraints and may result in a wide variety of designs and performances. Student-built cars that can travel a long distance are not the primary assessment criterion. Distance traveled and velocity are, however, the result of good use of experimental data in design. So you may choose to add extra points for performance as part of your complete assessment system.

## Notebook

Because process is an important objective of the unit, place a high priority on an organized, neat notebook. Keeping a notebook may be a new skill for the middle school student. Have students leave the first pages of their notebooks blank. A few days into the unit, ask students to make tables of contents for their notebooks during class time. You can do a quick assessment by asking a student to find something, based upon the table of contents. If you give quizzes, allow students to use their notebooks; this will help them appreciate what they have already learned.

### Design Portfolio

A second product of the unit is a student's design portfolio. It should include the goals and requirements, the prototyping process, sketches, specifications, and test results. A short writing assignment reflecting on the design process as a whole is a valuable component of this documentation. Some of the students' original ideas may first be recorded in their notebooks and then transferred into design portfolios at a later date.

## Oral Presentation

The final oral presentation emphasizes the cooperative nature of engineering. Make sure that the presenters invite questions. Establish the rule that if you (the teacher) raise your hand, presenters should call on you before addressing others who have questions. Points for the oral presentation might be discussed when you distribute the student handout, "The Mousetrap Car Challenge."

# Introduction:
# The Mousetrap Car Challenge

Your challenge is to build a car powered by a mousetrap. You will be part of a Vehicle Research Team (VRT). Your vehicle must be no longer than 45 cm (excluding the snapper arm) and no taller than 30 cm. To build your car, you will collect data from experiments on force and motion. You also will keep a notebook and a design port-folio. This handout should become the first page of your notebook. The chart on the following page, to be filled in throughout the unit, shows the products and activities that will make up your project.

When your car is built, your VRT will be expected to show the car to the class. In your two– to five–minute presentation, you will report your vehicle's characteristics, including mass and velocity. Be sure to include answers to the following questions in your final team presentation:

- How much force can your mousetrap provide?
- How is the force of the mousetrap transferred to the wheels of the car?
- What data did you use to choose your chassis, wheels, and bushings?
- How does the speed of your car compare to the speed of a bicycle?
- How does your car meet the design requirements?
- How did the design of your car change as you built it?
- How could your vehicle be improved?

# Unit 1: Introduction

## Unit Point Values and Grade Scale

| PORTFOLIO TABLE OF CONTENTS | DATE ASSIGNED | POINTS POSSIBLE | POINTS EARNED |
|---|---|---|---|
| 1 Unit Student Handout (this paper) | | | |
| 2 Design Portfolio Technical Paper | | | |
| 3 Assembled Design Portfolio | | | |
| 4 Team Project Presentation | | | |
| 5 Activity 1: Writing Design Briefs | | | |
| 6 Activity 2: Measuring Force | | | |
| 7 Activity 3: Writing a Design Process Paper | | | |
| 8 Activity 4: Testing Wheels and Axles | | | |
| 9 Activity 5: Measuring Lever-Arm Effects | | | |
| 10 Activity 6: Developing a Design | | | |
| 11 Activity 7: Constructing a Prototype | | | |
| 12 Activity 8: Measuring Motion | | | |
| 13 Notebook | | | |
| | TOTAL | | |

Performance Bonus (extra credit)

Grade Scale

GRADE

# Activity 1:
# Writing Design Briefs

*Question: What will we need to design a car powered by a mousetrap?*

## Procedure

In this exercise you will begin to describe the design of your mouse-trap car in a way that will help your team work together efficiently.

1. Watch your teacher's demonstration of the operation of a mouse-trap. The elastic material of the spring steel in the trap provides the force to catch a mouse. You can capture that force to power a car. Discuss with your team the rules you will need to make sure that no one is pinched by your mousetrap. Create a list of safety rules. Put them on the second page of your notebook.

2. Think about the problem. Begin by asking what a good mousetrap car should do. These are your goals. Engineers call these goals "product specifications." They may come from a contractor or from the marketing department of a company that employs an engineering firm. The class will discuss these goals together.

3. Put your goals on the third page of your notebook. Label them "Product Requirements." Here are some examples:

   - The car must roll.

   - The mousetrap must be connected to the wheels.

   - We must be able to cock the lever arm of the mousetrap.

   Some product requirements reflect the cost or difficulty of construction. For example, you might require that the car be built entirely from items that cost less than five dollars. This type of requirement would not necessarily affect the function of the design. Another requirement might be to build a car that could travel 100 kilometers an hour—but that would not be a realistic goal.

## Objectives

- To decide on the requirements of a successful vehicle

- To write a design brief that defines the major goals of the project

## Materials

- Copies of *Brainstorming, Mind Mapping,* and *The Design Process*

4. Write the design brief. This is a simple statement of what you have to do to make the best possible car. (This should not be more than one sentence and should summarize the requirements you have listed.) Discuss your sentence with your team before you write.

_____

_____

_____

_____

_____

_____

_____

5. Create a sketch that will allow you to visualize your ideas. If you get new ideas when the class brainstorms together, you can create another sketch. Do not erase or throw away your first one. Just number each sketch in order. When you have a sketch that your team agrees on, label all its parts.

6. Determine the design specifications for your sketch. These are the things you will need to do to make your own design work. The whole class may have the same requirements, but your team's specifications will be unique. When you agree, make a list of these specifications here:

_____

_____

_____

_____

_____

_____

_____

# Teacher's Guide to Activity 1:
# Writing Design Briefs

*In Activity 1 students use brainstorming and mind-mapping techniques to explore the requirements of the design problem; that is, those characteristics specified for the product. In the engineering world the requirements might be defined by the agency contracting a product (such as the National Air and Space Administration's requirements for a probe). Or requirements might be defined by a company's marketing department ("We need a cell phone smaller than a pack of cigarettes"). In the school environment these product requirements might be called goals.*

*Once the product requirements are established, engineers move on to brainstorming and development of prototypes. Today most prototyping is done by computer. In the classroom your students will begin with sketches. Students also will investigate the conditions necessary to build and test a prototype. These conditions are called product specifications, and they will differ for each team. Because science education is often based on finding the right answer, this stage may be your students' first opportunity to demonstrate their creativity.*

## Conceptual Development

This unit has many objectives that will be new for middle school students: distinguishing a product's design requirements from its specifications; brainstorming; developing a design in stages based on data acquisition, record keeping, and self-evaluation. You can support these goals by coaching and periodic discussions. *Brainstorming, Mind Mapping,* and *The Design Process* are supplementary student readings appropriate for Activity 1. You can assign them either at the outset or on alternate days.

Help students distinguish between design requirements and possible solutions in a class discussion. Use real-world examples from engineering. Use butcher paper or create transparencies so that you can re-post these requirements throughout the unit.

Although each team will have different solutions, try to develop class consensus on goals and requirements. Team solutions will be reflected in their drawings and their specifications. It will take some time for student teams to develop consensus on their specifications.

Explain to students that the design process is a process of inquiry and discovery—cyclical and evolutionary rather than linear in nature. Working drawings may be changed as new knowledge and insights emerge. Introduce a system through which students can number or date stages in the process. Encourage students to compare their sketches to the design requirements periodically. Provide input to keep students on track—it is easy for them to get far afield in this process. You might want to circulate among teams or have some teams leave their notebooks for checking each day.

Integrate your lesson with language arts. It is likely that students are learning process writing in language arts classes. Both design and writing may begin with brainstorming, but what an engineer calls a design might be called a rough draft in writing. If possible, use the same terms and help students relate language arts techniques to what they are doing in science.

## Extensions

Students could gain practice in writing design briefs to solve problems by referring to materials they are reading for their language arts classes. For example, in C. S. Lewis' *The Lion, the Witch and the Wardrobe*, the travelers have to cross the land of Narnia. This is a perilous journey. Students could write design briefs to create devices to keep the travelers safe.

Art that accompanies science fiction also can provide ideas. Although there are currently no vehicles that carry humans on interplanetary journeys, we understand the design requirements for such vehicles. Many artists have suggested possible prototypes. Comparing such drawings can help students understand that there are many solutions to a problem.

# Activity 2:
# Measuring Force

*Question: How do you measure the force exerted by a mousetrap?*

## Objectives

- To define force

- To measure the forces exerted by a mousetrap

- To find the relationship between the distance the trap is open and its force

## Materials

- Mousetrap

- 10-N spring scale

- Set of blocks, to hold the traps open at the desired distances

## Procedure

In this activity you will determine the force that can be produced by the steel of a mousetrap.

1. Explore the force of one newton (N). Pull gently on the spring scale. With your team think of some things that are approximately equal.

A force of 1 N feels about like _____

2. Explore the forces exerted by a mousetrap. Be very careful to follow safety rules. Mousetraps can hurt you—even break bones. Remember to take a zero reading of the spring scale before taking measurements. Make adjustments as necessary. Then hook the spring scale to the lever arm and lift gently. Watch the gauge as the lever arm gets higher. Be sure to hold the spring scale *perpendicular* to the mousetrap snapper arm when measuring force, as illustrated below. Discuss any measurement problems you have with your teacher before you begin to record data.

3. Measure the force the mousetrap exerts when it is closed. Hook the scale to the lever arm and lift *slightly*, just until you see the arm move. How much force does it take to just start the mousetrap? _____

4. Use each of the wooden blocks to hold the trap open. Pull the spring scale so that the lever arm just barely lifts up from the block, as shown below. Record the force measurement for each distance in the data table on the following page.

**Measuring Force of Mousetrap Using Blocks**

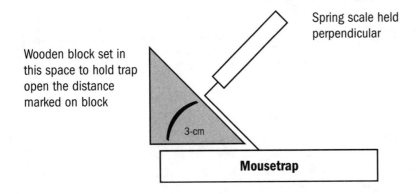

Spring scale held perpendicular

Wooden block set in this space to hold trap open the distance marked on block

3-cm

**Mousetrap**

5. After the force-distance measurements are recorded in the Force Measurement Data Table below, graph the data on the Force vs. Distance Graph. Distances measured (manipulated variable) go on the x-axis. Force (dependent variable) goes on the y-axis. Since distance is a continuous variable (and could be anything in between) the graph must be a line graph.

# Unit 1: Activity 2

## Force Measurement Data Table

| DISTANCE (cm) | FORCE (N) |
|---|---|
|  |  |
|  |  |
|  |  |
|  |  |
|  |  |
|  |  |
|  |  |
|  |  |
|  |  |
|  |  |

## Force vs. Distance Graph

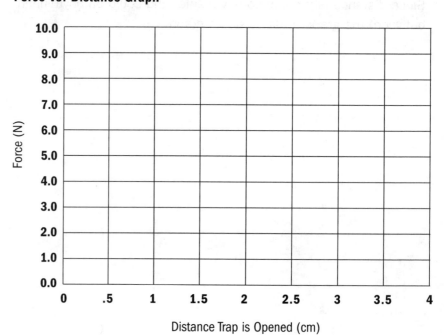

6. Discuss the Force vs. Distance Graph with your team. Then, in the space below, name and describe the relationship between the force exerted on the mousetrap and the distance by which it opened (as illustrated in your graph). Determine whether the relationship is **direct** (force increasing with distance) or **inverse** (force decreasing with distance).

_____

_____

_____

_____

_____

_____

_____

7. Discuss with your team how this relationship is useful in designing a mousetrap car. Record your team's ideas in the space below.

_____

_____

_____

_____

_____

_____

_____

# Teacher's Guide to Activity 2: Measuring Force

*Using a spring scale calibrated in newtons, students learn to measure the force of the snapper arm (lever arm) of a mousetrap. With the aid of blocks cut to hold open the mousetrap, students measure its force at several different positions. Because force will be affected by gravity, the scale must be perpendicular to the lever arm for each measurement. Students then graph the relationship between force and distance and hypothesize how to use this relationship in their design of a mousetrap car.*

## Conceptual Development

In this activity students learn how to measure force and how to collect and graph data. They name and describe a relationship between the two quantities measured. Although most students will not yet understand the mathematics, they are inductively discovering a direct linear relationship. Students may be able to guess the relationship correctly, but scientific process should be emphasized.

Graphing may require some special instruction. In mathematics, students often do one lesson on line graphs and one on histograms, but rarely do they discuss why one type or another is used. In this exercise the manipulated variable (distance) is continuous; that is, it could really be any number within the range. Therefore a line graph is appropriate.

## Extensions

Collaborate with the mathematics teacher in the graphing assignment, referring students to the appropriate lesson in the mathematics text for extensions.

The relationship between force and distance in mousetrap steel is an example of Hooke's law, discovered by Robert Hooke in England in the 17th century. This principle, which states that an elastic body bends or stretches in direct proportion to the force acting on it, applies to all elastic materials, including rubber and some kinds of plastics. Students may wish to research inventions that depend upon elasticity and their effects upon society in the Industrial Revolution.

Topic: Hooke's Law

Go to: *www.scilinks.org*

Code: FBP030

# Activity 3:
# Writing a Design Process Paper

*Question: How can you describe your design process for others?*

## Procedure

Scientists and engineers must communicate their ideas to one another and to people who depend upon their work. Although team members may exchange sketches, lists, and data, communication outside the team usually takes the form of clear informational writing.

1. Start your technical paper right now by writing a rough draft. (You can do this on paper or on a word processor.) In the rough draft try telling your story as if you were talking to a group of friends. Begin with the challenge, and then describe how the class set up the project requirements.

2. Set up titles for each section of your paper. (You can do this by putting a title on different sheets of paper or putting titles in your word processing file.) Titles should include: (a) Describe the Challenge (opening paragraph); (b) Define the Requirements; (c) Gather Information; (d) Explore Ideas; (e) Develop the Design; (f) Construct a Prototype; (g) Test and Evaluate; (h) Redesign; and (i) Conclusion (closing paragraph).

3. Start by writing the sentences for the second section. (The opening and closing paragraphs are best done last.) Your technical paper will have at least seven middle paragraphs, one for each phase of the design process. Each paragraph should describe what you did. You will probably revise the paper many times, so save your work carefully.

## Objectives

- To communicate to others the process used in design

## Materials

- Requirements list to be generated by the class

- Notes, sketches, and product specifications from your notebook

# Unit 1: Activity 3

## Writing Guidelines

**Describe the Challenge (opening paragraph):** Share some interesting or important details of the design challenge. Describe why you are doing the project and briefly tell what you want to accomplish.

**Define the Requirements:** Describe how the class decided on the goals and requirements for your process and what you did in Activity 1.

**Gather Information:** Describe how you gathered information for your project. Include your procedure and conclusions in Activity 2.

**Develop a Plan:** Describe how your team came up with your plan. Include how many sketches you did and how many ideas you had before you decided on the best one.

**Construct a Prototype:** Describe how you built your car. Include Activity 7.

**Test and Evaluate:** Describe how you tested your car. Include Activity 8.

**Redesign:** Describe the changes you made in the car after your tests and why you made them. Explain how these changes helped you meet the product requirements better.

**Conclusion (closing paragraph):** This should describe what you concluded and what you learned. You also can include what you would do differently next time.

# Teacher's Guide to Activity 3:
# Writing a Design Process Paper

*Students must be guided through the rough draft of their papers documenting their design process. Activity 3 gives them a format for each of the required nine paragraphs and advice on how to complete this task. This activity is ongoing; the rough draft should be revised periodically throughout the unit.*

## Conceptual Development

Students learn to write paragraphs and expository papers in middle school. This activity helps each student compose a technical paper documenting the design process. Students learn how each activity is associated with some phase of the design process and how the activities can be used as evidence of each phase. This activity should make writing the technical paper a manageable task.

Many students learn process writing in language arts. Stages outlined here are similar. Students also may learn to do rough drafts and revisions using word-processing equipment. This is the ideal way to encourage students to constantly improve their work. If students must use paper and pencil, encourage them to keep rough drafts in a separate binder or folder, so that each paragraph can be replaced as it is improved.

## Extensions

Newspapers often report on new inventions, but these reports occur after the invention is completed. Ask students to imagine they are writing a newspaper report of the activity inside an engineering facility (such as Microsoft inventing a new software product or an automobile company inventing a new car).

Technical writing is a specialized career field with attractive job opportunities. Students may wish to interview a technical writer or critique the manuals written for equipment by these professionals. One example is the manual that comes with your DVD or computer. Is it understandable? How are technical specifications translated into language that lay persons can understand? How could they be made better?

# Activity 4:
# Testing Wheels and Axles

*Question: How do you select axle materials to make your vehicle roll easily?*

## Objectives

- To test a variety of materials for axles in order to find materials that create the least friction

- To test a variety of materials for axle bushings in order to find the best combination for vehicle performance and endurance

## Materials

- Four axle rods (wood, plastic, and metal—all about 0.5 cm in diameter and 12 cm in length) per team

- Small-diameter plastic drinking straws, cut to 10 cm in length

- Two sections of brass tubing (10 cm in length with an inside diameter of 6 mm) per team

- Four car wheels per team

- Test block made of pine (with 2.5-cm x 10-cm cross section and a length of 15 cm, equipped with holes for axles and bushings) per team

*continued on next page*

## Procedure

One of the most important considerations in your mousetrap-powered vehicle design is the wheel-and-axle assembly. You will have to decide on wheels, axles, and axle bushings. A wood axle turning in a hole drilled through wood produces a lot of friction. A bushing is a tube used to reduce the amount of friction on the axle. You should decide which (if any) bushing would reduce friction best. You also will have to decide on the size of your wheels and on how you will attach the wheels to the axles.

The axle blocks have four holes in them. The front axle will have a plastic bushing (a straw) in it and will remain constant for all the tests. The three rear-axle holes will be the test holes. Two of the test holes will contain bushings. You will use a metal rod for the front axle. For the three test axles, one will be made of wood, one of plastic, and the other of metal. There will be nine tests altogether, with three different choices of bushings and three different types of axles.

1. Review your design requirements. Your vehicle will be judged on how far it travels. Remember that the size of the wheels will not be the only decision. Bigger wheels may go farther but may have more friction.

2. Slide a plastic bushing and metal axle through the front hole and attach the wheels. Cut or select a wood axle, place it in another wood hole, and attach the wheels. The assembly is now ready for testing.

3. Conduct the test by having the test block (with wheels and axles attached) run down a ramp. Attach a piece of cardboard to the end of the ramp to make a smooth transition from the ramp to the floor. Carefully measure how far the assembly travels, and record the result.

_____

_____

_____

_____

_____

4. Conduct the following tests using different axle materials with the different bushings. Keep the front wheel the same for all the tests.

    Test 2: wood axle, plastic bushing

    Test 3: wood axle, brass bushing

    Test 4: metal axle, no bushing

    Test 5: metal axle, plastic bushing

    Test 6: metal axle, brass bushing

    Test 7: plastic axle, no bushing

    Test 8: plastic axle, plastic bushing

    Test 9: plastic axle, brass bushing

Record your results in your notebook.

_continued from previous page_

- One 61-cm x 244-cm piece of plywood (or access to a school ramp)

- Cardboard to stop the test block

- Hacksaw

- Meterstick

- Eye protection

# Unit 1: Activity 4

5. Evaluate your results with your team and write a conclusion.

# Teacher's Guide to Activity 4:
# Testing Wheels and Axles

*Students are given a wheel-and-axle test block that can accommo-date a variety of materials for bushings. They also are given axles made from a variety of materials. Students follow test procedures that aid in selecting the best combination of axle and bushing mate-rial. Testing can be accomplished on a plywood ramp or on a ramp outside the school. Students then utilize this information when designing the wheel-and-axle assemblies for the prototype.*

## Conceptual Development

This experiment requires students to evaluate and compare two different variables simultaneously. For most students this will be a difficult process. You can help them work through the evaluation by asking questions and giving plenty of time to explain the logic of their processes. Combinatorial operations—such as the process involved in Activity 4—is a stage of logical thinking that develops in middle school.

In performing the tests in Activity 4, students learn about friction, a force that opposes the turning of the wheel. Students will experi-ment to determine the combination of axle and bushing materials that best meets the design specifications. They will use this infor-mation to determine which axle housing creates the least friction. As testing continues, have students enter a definition of friction in their notebooks in the glossary section. (Unit 2, "The $CO_2$ Car" provides more empirical studies of friction.)

Safety is an important priority in this and all technology labs. The test blocks must be fabricated by an adult. If you allow students to cut their own axles with a hacksaw, review safety procedures and require eye protection. Set up a cutting station at the front of the room where you can constantly supervise.

Topic: friction

Go to: *www.scilinks.org*

Code: FBP037

## Extensions

Another test procedure might be to investigate the amount of force required at the axle to turn wheels of different sizes. Have a variety of materials (such as hooks and small nails) that students could use to attach the axle to the vehicle chassis instead of using the bushings.

Students may wish to investigate and compare advertisements for automobile tires. Some tires get far better mileage than others.

**Axle-and-Bushing Test Block**

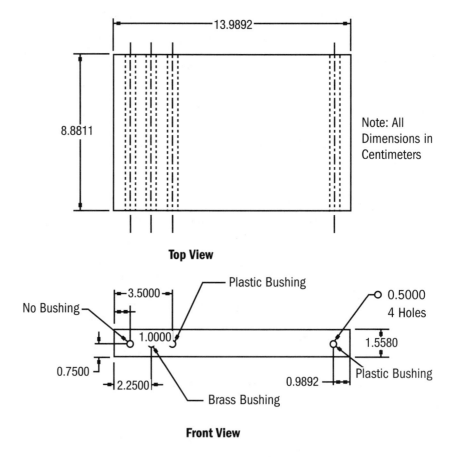

13.9892

8.8811

Note: All Dimensions in Centimeters

**Top View**

Plastic Bushing

No Bushing

3.5000

1.0000

0.5000
4 Holes

1.5580

0.7500

2.2500

0.9892

Plastic Bushing

Brass Bushing

**Front View**

# Activity 5:
# Measuring Lever-Arm Effects

*Question: What happens to the force exerted by a mousetrap when the snapper (lever) arm is made longer?*

## Procedure

Review your results for Activity 2. Even though a mousetrap is designed to exert a large applied force, you will be measuring a very small force with sensitive spring scales, so be very careful not to harm the scales by pulling them too hard.

1. To measure the effect of a lever arm, extend the mousetrap snapper arm by attaching a long, thin, strong piece of wood to it. You will use the length of the normal lever on a mousetrap (l) as your unit of measurement. Measuring the force at distances of l, 2 x l, 3 x l, 4 x l, etc. will give a clear picture of the relationship between the magnitude of the force and the length of the lever arm. Measure the length of the trap's arm in centimeters. Mark your piece of wood in these units: 2 l, 3 l, 4 l, etc.

**Measuring Force Distances**

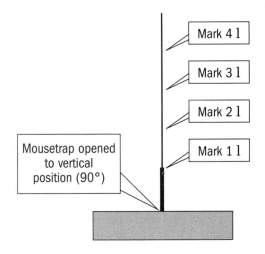

Mark 4 l

Mark 3 l

Mark 2 l

Mark 1 l

Mousetrap opened to vertical position (90°)

## Objectives

- To find out how the lever arm transforms force

- To measure the forces exerted by a mousetrap with a longer lever arm

- To graph the relationship between force and the length of the lever arm

- To describe how a lever trades distance for force

## Materials

- Mousetrap

- 10-N spring scale and 2.5-N spring scale

- Strong piece of material, to extend the snapper (lever) arm of the mousetrap

# Unit 1: Activity 5

2. Hold the lever arm straight up (in the vertical position). Be sure to keep the spring scale perpendicular to the lever arm, and be consistent. Record the pull force at each mark on the Force vs. Distance Data Table below. Make a line graph of your results in the Force vs. Lever-Arm Length graph.

**Force vs. Distance Data Table**

| DISTANCE (marks) | FORCE (N) |
|---|---|
|  |  |
|  |  |
|  |  |
|  |  |
|  |  |

**Force vs. Lever-Arm Length Graph**

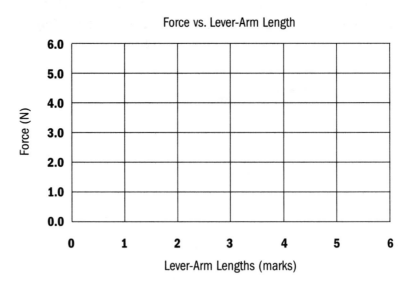

3. Discuss with your team what is revealed in the graph, so that you can name and describe the relationship between the mousetrap force and the length of the lever arm.

4. How can the relationship between the force and the lever arm be used in the design of a mousetrap car? Record your thoughts below.

_____

_____

_____

_____

_____

# Teacher's Guide to Activity 5:
# Measuring Lever-Arm Effects

*Using a strong extension to the snapper lever arm of a mousetrap, students measure the force exerted at several different points with the mousetrap in its 90-degree position. Students graph this relationship and discuss how this relates to their mousetrap car. They discover that a longer arm makes the extension (launch) of the vehicle easier. However, the additional mass has a negative relationship to velocity.*

*In this exercise students use a unit equal to the length of the lever arm (l) rather than centimeters to measure the relationship between distance and force. This unit may confuse students. Take time to discuss the origin of units and why a longer unit produces clearer results in this experiment.*

## Conceptual Development

A lever arm is a simple machine that transforms forces. A lever can convert a large force that acts over a short distance into a small force that acts over a large distance and vice versa. This force-distance trade-off is the principle behind all simple machines. The inverse relationship between force and distance is easily measured, especially when the units of distance are multiples of the length of a lever arm: twice the length, three times the length, etc. But the inverse relationship between force and distance can be difficult for students to grasp.

The inverse relationship revealed in this experiment is important to the mousetrap car because a longer arm can increase the force applied to the car. But, of course, there is a limit to the advantage that this can produce because it increases the mass of the car.

SCI LINKS
THE WORLD'S A CLICK AWAY

Topic: levers

Go to: *www.scilinks.org*

Code: FBP042

**Data and Graph for Force vs. Lever-Arm Length of Victor™ Brand Mousetrap**

**Force vs. Distance Data Table**

| DISTANCE (marks) | FORCE (N) |
|---|---|
| 1l | 5.4 |
| 2l | 2.7 |
| 3l | 1.4 |
| 4l | 0.7 |
| 5l | 0.3 |
| 6l | 0.2 |

**Force vs. Lever-Arm Length Graph**

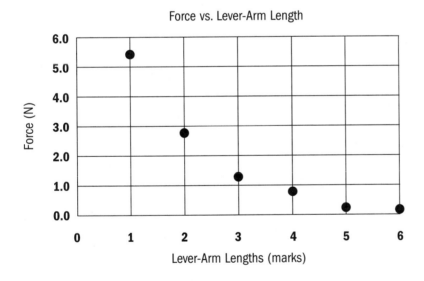

## Extensions

Challenge students to find lever arms employed in the classroom. Lever arms are everywhere. Ask students how much force they need to exert to open your classroom door. Then ask them if it matters where they push and why. A lever-arm scavenger hunt around school or in their homes is another activity students might enjoy.

# Activity 6:
# Developing a Design

*Question: How do you describe design ideas and communicate your design to others?*

## Objectives

- To use sketching as a tool to explore a variety of ideas

- To use sketching as a tool to work out design details

- To use sketching as a tool to communicate design ideas to others

## Materials

- Copy of *Sketching Guide*

- Several sheets of grid paper

- Colored pencils

## Procedure

1. Begin by completing the activities in the *Sketching Guide*. These won't take long, and they are fun. Practice making a few sketches before working on your vehicle design. Remember that you are only communicating with your team, so don't worry about how it looks! You may make your sketches on blank sketch paper (no lines) or on grid paper. The paper can be three-hole punched and added to your notebook later.

2. Begin exploring possible designs by making sketches. These can be either two-dimensional or three-dimensional in appearance. You need not show everything in a single sketch, but you may wish to use detailed sketches to show the various parts of the vehicle that are too small to be seen in the overall sketch.

3. Continue making sketches until you run out of time. There is never enough time to explore all possibilities! It is important to explore as many as you can in the time allowed.

4. After you have explored many ideas, you need to pick the best one and develop it so that it can be used in constructing your mousetrap car. Once your reach this stage, make drawings showing the top and side views of the vehicle.

5. Use colored pencils to show the various parts of the vehicle.

6. Include the dimensions of the various parts of your vehicle on your drawings, so that you can build the vehicle to the proper size. This will communicate design ideas and goals and is necessary for design approval.

# Teacher's Guide to Activity 6:
# Developing a Design

*Students learn to use paper and pencil to pursue a variety of ideas through the use of simple sketches. They learn some simple technical sketching techniques, including the use of grid paper, both isometric and orthographic. The use of color is encouraged, to aid students in identifying different parts and materials. Color is also a good motivational tool.*

## Conceptual Development

Through simple, freehand sketches, students begin to develop the cognitive processes required to form visual images in order to get ideas down on paper. The ability to make competent freehand sketches is a skill that can be learned—but cannot be developed quickly or easily. It is not simply a matter of developing psychomotor skills, as in the use of tools. Students must learn to visualize in the ways that artists do. This is a long and complex process. The following books may help you get a better understanding of the development of freehand sketching skill:

Edwards, Betty. *New Drawing on the Right Side of the Brain.* London: HarperCollins, 2001.

Raudebaugh, Robert A. *Visualization, Sketching and Freehand Drawing for Engineering Design.* Mission, KS: Schroff Development Corporation, 1999.

It is important to realize that developing drawing skills requires effort and practice. These skills will not come through these few drawing activities. This is why sketching is included in all of the design-and-build activities throughout this book. Students should not expect instant success. Make sketching a daily classroom routine. Suggest short exercises during the time used to take attendance. Have students retain their sketches, so that they can see improvement over time.

## Extensions

Computer-assisted design (CAD) is a valuable skill in today's technological world. Many students may have the skill to do their drawings in a CAD or mechanical drawing program on the computer. This is a skill that should be encouraged. Students who are interested in pursuing drawing beyond this course could peruse *Visualization* materials available on CD-ROM or visit Schroff Development Corporation's website at *www.schroff.com*.

# Activity 7:
# Constructing a Prototype

*Question: What is the best way to construct the prototype of the design?*

## Procedure

In this activity you will build your prototype vehicle. With your team you will have to decide how to attach the wheels and axles to the chassis of the vehicle, transfer force from the spring to the wheels and axles, align the wheels and axles so that the vehicle travels on a straight path, and assemble all of the parts to finish making the vehicle. Use data from your experiments to help make team decisions.

1. Plan a course of action. Make a list of the steps you will take. It might look something like this:

   - Measure material for chassis and make cuts.
   - Cut axles to proper length at the hand tool station.
   - Drill holes for axles (if necessary) at the hand tool station.
   - Assemble chassis.
   - Attach mousetrap to chassis.
   - Attach lever arm (if any) to mousetrap.
   - Attach wheels to axles in axle housing.
   - Attach wheels and axle-housing assemblies to chassis, or insert axles into axle holes in chassis and attach wheels.
   - Attach mechanism for transferring force in mousetrap spring to axles.

2. Get the course of action approved by your instructor. Among other things, your plan must reflect the design of your vehicle.

3. During ALL phases of prototype construction, safety glasses MUST be worn.

## Objectives

- To select the best way to assemble the mousetrap vehicle

- To develop efficient construction methods

## Materials

- Safety goggles

- Assorted discs (CDs, old records, lids, etc.) for wheels

- Material selected for the axles and bushings

- Mousetrap for each team

- String (75 cm)

- Material to be used in creating the vehicle chassis (the frame that holds the mousetrap and the wheels): wood scraps (particularly thin material), tongue depressors, strips of plastic, metal rods (coat hangers, welding rods)

- Material for lever-arm extension (optional)

- Hand tools: hacksaw, hand drill, hammer, small nails

- Wood glue

# Unit 1: Activity 7

4. Carry out each step of the course of action until the vehicle is complete. You may want to experiment with methods of fastening the material to find the method that best holds the vehicle together. It will have to undergo several tests.

5. Once your have completed all the steps of your planned course of action, test the vehicle to determine whether it works as intended. Make any modifications that may be necessary in order to get the car working.

6. Record each step of your procedure in your notebook. Do not forget to record mistakes too.

7. Test your vehicle. Make a table to record how far it travels in each test run. Include the table in your notebook.

NATIONAL SCIENCE TEACHERS ASSOCIATION

# Teacher's Guide to Activity 7:
# Constructing a Prototype

*Students embark on the task of building the cars they have designed. They will be confronted with the reality of things that look good on paper but do not actually work. They will learn that it is important to test, experiment, and redesign aspects of their vehicle as they go. They also will learn that attention to detail often will have a significant effect on quality—both on the performance and the endurance of their vehicles. Make sure to establish discipline and safety rules in advance.*

## Conceptual Development

Translating their designs into reality, students begin to learn about the interaction between tools and materials, including methods of fastening materials to make up configurations and assemblies and the importance of experimentation and testing. It is important at this point to emphasize the value of the scientific experiments students have conducted and to refer them to their data when decisions must be made.

Monitor use of tools carefully. No power tools should be used. If you allow students to use hacksaws and hand drills in the room, set up a single station for this activity where you can watch carefully. Insist on eye protection for all sawing and hammering. If you use glues, ensure adequate ventilation and remember that Materials Safety Data Sheets must be available at all times.

## Extensions

Have students research the periodic redesign of vehicles by automobile manufacturers. For example, the periodic redesign of the Ford Mustang occurred over a twenty-year period. In contrast, vehicles such as the Volkswagen Beetle remained the same for periods of eight to ten years. Ask students why vehicles are redesigned. (The answer is not always greater efficiency or safety. Often redesign is related to changing consumer tastes and needs, which are translated as requirements from marketing departments.)

# Activity 8:
# Measuring Motion

*Question: How do you measure the motion of a mousetrap car?*

## Objectives

- To measure and graph motion in terms of distance and time

- To calculate average speed from distance-time data

- To describe motion using the concepts of distance, time, and velocity

## Materials

- Metersticks

- Stopwatch

- Tape for marking track and meter intervals

- Completed mousetrap vehicles

## Procedure

When your mousetrap vehicle is completed and can travel at least five meters reliably, you will measure its velocity. Your VRT needs to be coordinated for this activity. Everyone should have a role. Decide on the jobs in advance. Respect other teams who are recording data. Do not confuse them by shouting out the times.

1. Prepare a flat, straight track at least 10 m long. Mark off distances of 1 m, 2 m, 3 m, etc. with a piece of tape. Label the tape accordingly, and be sure to remove it from the floor when you have finished. Practice running the mousetrap car down the track.

2. Run the car down the track, and record each time the car hits a meter mark. Repeat this procedure three times to insure greater accuracy. Record the time for each meter on the Time Trial Data Table. Average the times for the three trials and enter the times on the Average Time Data Table.

## Data and Graph for Force vs. Lever-Arm Length of Victor™ Brand Mousetrap

### Time Trial Data Table

| DISTANCE INTERVALS (m) | TIME TRIAL (s) | TIME TRIAL 2 (s) | TIME TRIAL (s) |
|---|---|---|---|
| 1 | | | |
| 2 | | | |
| 3 | | | |
| 4 | | | |
| 5 | | | |
| 6 | | | |
| 7 | | | |
| 8 | | | |
| 9 | | | |
| 10 | | | |

### Average Time Data Table

| DISTANCE INTERVAL (m) | AVERAGE TIME (s) |
|---|---|
| 1 | |
| 2 | |
| 3 | |
| 4 | |
| 5 | |
| 6 | |
| 7 | |
| 8 | |
| 9 | |
| 10 | |

# Unit 1: Activity 8

3. Graph the results of the trials in Distance vs.Time Graph below.

**Distance vs. Time Graph**

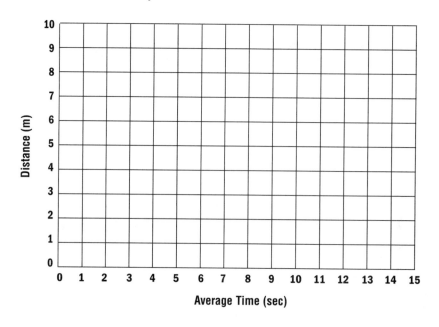

4. Discuss with your team the relationship of distance to time in the graph. Is it direct or inverse? Does the car take the same time to cover each meter? Record the nature of the distance/time relationship in the space below.

_____

_____

_____

_____

_____

_____

_____

5. Calculate the average velocity of the car. (Divide the distance traveled by the time taken to travel that distance.) Show your work, and be sure to express the velocity in units of m/s. (The velocity is positive because the car is moving away from the starting point.)

_____

_____

_____

_____

_____

6. A velocity of 1 m/s range is the rate of a very fast walk. To find out the speed of your mousetrap car in km/hr, use the conversion factor 1 m/s = 3.6 km/hr.

_____

_____

_____

_____

7. Describe the motion of the mousetrap car over the 10-m track in complete sentences, using the words "distance," "time," and "velocity."

_____

_____

_____

_____

_____

# Teacher's Guide to Activity 8: Measuring Motion

*After students have built their mousetrap cars, they measure motion by recording times as the cars pass marks along the floor spaced one meter apart. This activity works well if their cars move at about the speed of walking. If cars travel faster than this, some other means of collecting data may be necessary. The purpose of measuring time at each meter interval (rather than the total time divided by total meters) is so that students can see the acceleration due to the force of the mousetrap and the gradual deceleration of friction.*

*Students may use stopwatches, wristwatches, or probeware to measure time. After displaying their data on a graph, students calculate the average velocity of their cars in several units. They then write a description of their motion, using the words "distance," "time," and "velocity."*

## Conceptual Development

This activity may be the first time that the students have measured velocity. Make sure that they understand rate and time and that they enter appropriate definitions in their glossaries. A secondary purpose is for students to discover that repeated measurements increase accuracy.

Velocity is a vector quantity, which has both speed and direction. To avoid misconceptions later, emphasize direction in the language you use to describe the vehicle's motion: "It is moving ___km/hr away from the point from which it started."

Velocity changes throughout the run; the vehicles move quickly when snapped and slow down due to the force of friction.

In an actual engineering project the motion of the prototype would be measured with computerized sensors. This probeware is commonly available at a reasonable price and may be accessible in your middle school. Using motion detectors, the deceleration of friction is easy to quantify.

Topic: velocity

Go to: *www.scilinks.org*

Code: FBP054

## Extensions

Begin exploring automobile safety with your students. Ask them to consider how far they move each minute and second while traveling in a car on a freeway. (The answer is roughly 30 meters each second.) As the inevitable crashes occur with the mousetrap vehicles, you can easily remark on the force of the collisions. This will prepare students for later studies of momentum.

Then extend the lesson to show students the effects of the cars' momentum. Put an egg on top of one of the cars (preferably, your own model) and allow it to crash into a barrier. Then create a short barrier that blocks only the bottom of the vehicle. Place a doll or stuffed figure on the car and allow it to crash, sending the figure flying forward. Students may want to videotape these demonstrations to create "commercials" for seatbelt use for your school's video system.

# Teacher's Guide to Unit 2:
## The $CO_2$ Car

*In this unit students create a vehicle that moves quickly on a track, using $CO_2$ as a propellant. Students learn how mass, shape, wheel alignment, and frictional force affect the motion of the $CO_2$ car as they learn how to work within specific design requirements. The design process helps students develop prototyping skills. After completing the cars, students test them using specified performance criteria. Then, using mass vs. velocity data, students redesign their cars to improve performance. The final activity is a Great Race that allows students to compare the performance of various cars built to the same requirements.*

## Overview

This unit provides background for extending lessons on automobile safety into new areas. The high speed of the $CO_2$ vehicles introduces the issue of control into the design process. Students examine wheel alignment to reduce the potential of rollover in model trucks. The unit also introduces the concept of center of gravity. Using force diagrams, students can examine the differences among vehicle designs and compare the stability of various kinds of vehicles under high-speed conditions in a structured laboratory situation.

This guide suggests using commercial kits for $CO_2$ vehicles. If your facilities permit, you may wish to provide students with wooden blanks and allow them to construct their own cars. Partially shaped wooden blocks are also available in hobby shops. Shaping the cars requires the use of more tools and much more stringent safety precautions. (See figure on the following page for the dimensions for these blocks if you choose to use them.) The lessons on mass, velocity, and friction can be achieved with either commercial or student-built car bodies.

## Materials

- Copy of student handout, "The $CO_2$ Car Challenge," for each student

- Copy of student handout, "Prototype Manufacturing Plan," for each student

- Copies of each of the eight activities, plus *Sketching Guide* and *The Manufacturing Process* for each student

- One $CO_2$ car kit per team, obtained from vendors such as the following:

  a) Pitsco, P.O. Box 1708, Pittsburg, KS 66762; 1-800-835-0686; *www. pitsco.com*

  b) Kelvin World, 10 Hub Dr., Melville, NY 11747; 1-800-756-1025; *www.kelvin.com*

  c) IASCO, 5724 W. 36th St., Minneapolis, MN; 55416-2494; 1-888-919-0899; *www.iasco-tesco.com*

  d) Car bodies also can be obtained from Pinewood Derby products or hobby stores.

*continued on next page*

### Materials *continued*

- One toy truck per team (Sample truck used is 18 cm long, 8 cm wide, and 8 cm tall, Strombecker® 1958 Ford)

- Four 250-gram masses, also used as 2.45-N weights, per team

- One balance per team

- One 2.5-N spring scale per team

- One meterstick per team

- One stopwatch per team (accurate to .001 s)

- One centimeter ruler per team

- One $CO_2$ car race track (see vendor list above)

- Hand tools

- Three $CO_2$ cartridges per team

- Lubricants (graphite)*

- Acrylic paints*

- Glues*

- Optional: Individual car blanks (See figure for dimensions.)

* Materials Safety Data Sheets required

Topic: $CO_2$

Go to: *www.scilinks.org*

Code: FBP058

### Dimensions of Blank for $CO_2$ Car Body

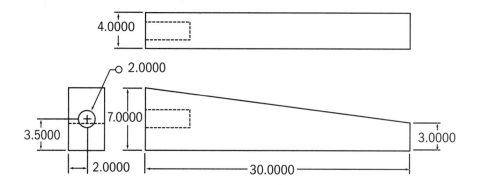

Note: All dimensions in centimeters

If you did not cover the unit on the mousetrap car, you will need to adapt two activities from that unit—"Activity 1: Writing Design Briefs" and "Activity 3: Writing a Design Process Paper"—for use in the $CO_2$ car unit. Take some time to review the concepts in those activities to make the most of the $CO_2$ vehicle design experience.

## Conceptual Development

To launch the unit, introduce the challenge of designing and building a $CO_2$ car by distributing the student handout, "The $CO_2$ Car Challenge." The handout gives students an overview of expectations and a table for recording points earned throughout the unit. Tell students to make the handout the first page of their notebooks.

Mass, force, and energy are difficult concepts for students. In this unit students move from thinking of mass as "an amount of matter" to thinking of mass as "the property of matter that resists changes in motion." They investigate inertia and Newton's laws of motion. Because the work is done in metric units, distinguishing mass from weight (the force of gravity) is simplified.

Middle schoolers intuitively think of forces as pushes or pulls. The next step is to convince them that force is needed to stop or change the direction of movement (inertia) or that forces are balanced in nature. Throughout this unit it will be important to monitor the development of understanding and potential misconceptions through continued dialogue with students. Activity 3 also introduces the use of vector diagrams to communicate ideas about force.

Models are physical or mathematical representations of scientific ideas. As they explore forces, students begin to use vector diagrams as models. Physics conventions represent forces as arrows, with the arrow length representing the magnitude of force in a specified direction. Using careful sketching techniques, students can develop a good basic understanding of vector forces without trigonometry. Activity 3 provides a set of steps for these graphic representations. It is important to monitor this process carefully so that students do not develop misconceptions that would make later work in physics more difficult. Vector diagrams can be used throughout the unit as embedded assessments of conceptual development. The diagram below gives an overview of the eight instructional activities and how they link to the design process.

## $CO_2$ Car Challenge
**Design Process With Associated Instructional Activities**

DEFINE THE PROBLEM
Activity 1: Requirements and Constraints

GATHER INFORMATION
Activity 2: Mass and Motion
Activity 3: Force Diagrams
Activity 4: Frictional Force

EXPLORE IDEAS
Activity 5: Writing Design Briefs

DEVELOP THE DESIGN
Activity 5: Writing Design Briefs

CONSTRUCT PROTOTYPE
Activity 6: Prototype Development

TEST AND EVALUATE
Activity 8: Time and Velocity

REDESIGN
Activity 7: Performance Testing

**Team Project Presentation**

## Tips for Making the Unit Work

*Arrange the classroom for movement and active learning.* Space is an important factor for safety. Establish patterns of movement within the room. If you have access to a technology lab, use it for the construction process.

*Interact with each student frequently.* Monitor force diagrams carefully. To test conceptual understanding of the significance of the arrows, have students explain what the arrows represent. Ask challenging questions throughout the process.

*Communicate high expectations.* The activities in this unit require very different skills: physical coordination, conceptual development, visualization, and organization. Each student will have different strengths and areas to improve. Encourage team members to recognize one another's abilities.

*Give praise and encouragement.* Be conscious of students who may be struggling and praise them when they make progress. Find at least one area to praise for every student.

*Communicate affection and regard.* Support a wide range of skills by complimenting students on what they can do. Humor and positive feedback will increase achievement.

*Emphasize organization.* Students should maintain notebooks of definitions and activities. The organization and table of contents of these notebooks should be their responsibility. Information from the notebooks will become part of the design portfolio and final presentations.

*Give students responsibility.* Rotate clean-up tasks among members of the class on a regular basis throughout the course of this unit. Involve them in the display of work in progress as well as in the results of completed activities.

*Treat all students equally and fairly.* Do not permit more able students to do tasks for others. Encourage students to tutor and advise one another, but insist that all students carry out the tasks that they are physically capable of doing.

## Physics Background

This unit builds on the mousetrap car activities in Unit 1. Check the conceptual development section of the teacher's guide to Unit 1 to be certain you have covered the concepts fundamental to Unit 2. As you introduce each of the instructional activities, you may need to present some of this material in class. Students should maintain their own glossaries of terms and ideas, such as the basic terms listed below, in their notebooks. It may also be helpful to review pertinent points as you complete each activity.

**Mass (m):** the "amount of material" in an object, measured in kilograms (kg). Mass controls an object's motion.

**Friction ($F_f$):** the force that opposes the motion of a body over or through another, measured in newtons (N). The coefficient of friction for two sliding materials is obtained by dividing the force of friction by the force of gravity.

**Weight ($F_g$):** the force caused by the gravitational attraction between Earth and the object, measured in newtons (N). (On Earth 1 kilogram weighs about 9.8 newtons.)

**Normal Force ($F_N$):** the force exerted perpendicular to an object by a supporting surface (such as a floor, a ramp, or a sidewalk). In this unit the normal force acts on the wheels of a vehicle.

**Center of Gravity:** point at the center of an object's weight distribution where the force of gravity can be considered to act. An object can be balanced at its center of gravity.

## Time Management

A suggested schedule for covering all the activities of this unit over the course of 20 school days (four five-day weeks) is given in the table below. It is based on allotting one class period of 45 minutes to the unit per day.

### Time Management for $CO_2$ Car Unit

| DAY NUMBER | ACTIVITY |
| --- | --- |
| 1 | $CO_2$ Car Challenge Unit Overview |
| 2 | Activity 1: Design Requirements and Constraints |
| 3 | Activity 2: Mass and Motion |
| 4 | Finish Act 2: Mass and Motion |
| 5 | Activity 3: Force Diagrams |
| 6 | Finish Activity 3: Force Diagrams |
| 7 | Activity 4: Frictional Force |
| 8 | Finish Activity 4: Frictional Force |
| 9 | Activity 5: Writing Design Briefs |
| 10 | Finish Activity 5: Writing Design Briefs |
| 11 | Activity 6: Prototype Development |
| 12 | Design and Build $CO_2$ Car |
| 13 | Design and Build $CO_2$ Car |
| 14 | Design and Build $CO_2$ Car |
| 15 | Activity 7: Performance Testing |
| 16 | Activity 8: Time and Velocity |
| 17 | Portfolio Preparation |
| 18 | $CO_2$ Car Great Race |
| 19 | Portfolio and Presentation; Preparation |
| 20 | Presentations; Portfolios Due |

## Unit Assessment Criteria

There may be no need to give pencil-and-paper tests to assess students' understanding and performance; completion of the project, together with the design portfolio, are good evidence of learning. Request that students respond to questions with vector (force) diagrams. If you choose to add objective questions, include diagrams, data, and real-world situations to make assessments more authentic.

## Notebook

Keeping an organized, neat, chronological set of classroom notes in binders may be a new skill for middle school students. Have students make their own tables of contents for the contents of their notebooks. Quick assessments can involve asking students to find something in the notebooks using the table of contents. Take a close look at a few notebooks each day.

## Design Portfolio

The design portfolio is a formal communication of the ideas originally recorded in the notebooks. It should be a polished, attractive collection of drawings and data, combined with a short expository essay on the design process.

## Team Presentation

The Great Race is a thrilling event, even though building the fastest car is not the goal of the project. Although the $CO_2$ car design-build project has many constraints, students manage to build cars that perform quite differently. To determine the winner(s) of the race, you should establish some objective performance criteria that most students can achieve, such as being among the cars that finish within 10 percent of the shortest time or the greatest average velocity. Other considerations might include stability, looks, and ease of manufacture.

# Introduction:
# The CO₂ Car Challenge

Your challenge is to design and build a car propelled by $CO_2$. The design must enable your car to travel a track as quickly as possible. You may use only the materials provided, and your vehicle must remain on the track for the entire run.

You will be a member of a Vehicle Research Team (VRT). The engineering process is very important. With your team you will complete eight activities. During the activities you must keep an organized notebook containing data, definitions, and background information. This handout, "The $CO_2$ Car Challenge," should become the first page of your notebook. At the end of the project, you will produce a design portfolio. This is a technical paper that describes the process you used and includes your sketches and performance data. The chart on the next page, to be filled in throughout the unit, shows the products and activities that will make up your project.

At the end of the project your Vehicle Research Team (VRT) will make a short presentation about your $CO_2$ car to the class. In your final presentations, keep the following points in mind:

- The team presentation should show the actual car.

- You should show how the design changed.

- All team members must participate equally in the presentation.

- You must explain how your research data influenced your design.

- The team should explain how the design could be improved.

**Unit Point Values and Grade Scale**

| PORTFOLIO TABLE OF CONTENTS | DATE ASSIGNED | PERCENT OF GRADE | POINTS EARNED |
|---|---|---|---|
| 1  Unit Student Handout (this paper) | | | |
| 2  Design Portfolio | | | |
| 3  Assembled Design Portfolio | | | |
| 4  Team Project Presentation | | | |
| 5  Activity 1: Design Requirements and Constraints | | | |
| 6  Activity 2: Mass and Motion | | | |
| 7  Activity 3: Force Diagrams | | | |
| 8  Activity 4: Frictional Force | | | |
| 9  Activity 5: Writing Design Briefs | | | |
| 10  Activity 6: Prototype Development | | | |
| 11  Activity 7: Performance Testing | | | |
| 12  Activity 8: Time and Velocity | | | |
| 13  Notebook | | | |
| TOTAL | | | |

Performance Bonus (extra credit)

Grade Scale

GRADE

# Activity 1: Design Requirements and Constraints

*Question: How can you design and build a fast car powered by $CO_2$?*

## Objectives

- To identify design requirements and constraints

- To explore as many designs as possible within the constraints

- To identify how to build a $CO_2$ car that is safe

## Materials

- $CO_2$ car kit

## Procedure

Before you begin this design activity, take a few minutes to think about safety. The car will be traveling down a track at high speed. Could anything happen that would be hazardous? What about the construction of the car? When it hits the end of the track, is there any danger that it might fly apart? You will have to test run the car several times. If it breaks before the tests are over, you will not be able to complete all the activities of this unit or to participate in the Great Race. What about keeping it on the track? Is that a concern?

The handling of a car is a serious concern to designers. Safety is one of the most important factors to an automotive engineer. Safety cannot be compromised for appearance or performance. It is important for you to begin thinking about these issues now in the early stages of the design process.

1. Re-read the design challenge, to familiarize yourself with the design requirements for the $CO_2$ car. What is your goal?

_____

List the design requirements on the second page of your notebook. Include safety factors.

2. Inventory the parts in your kit. Write down the function of each part and what it does.

Car body _____

Metal axle _____

Bushing _____

Wheels _____

Hook eyes _____

Cartridge _____

3. To run, each part of the car must perform well. After each component write a sentence explaining what the best possible performance would be:

Car body _____

Metal axle _____

Bushing _____

Wheels _____

Hook eyes _____

Cartridge _____

4. You can use only these parts. But they can be modified in different ways. Begin by brainstorming with your team members about the variables you could change. You will be able to refine your ideas with investigations on mass and motion and on force diagrams later. List some brainstorming ideas in your notebook.

5. Write down in your notebook a list of steps you could take that would enable you to do the best job of meeting the design specifications for your $CO_2$ car.

# Teacher's Guide to Activity 1: Design Requirements and Constraints

*Students inventory the parts within the $CO_2$ car kit and identify how each part affects the performance of the vehicle. Using brainstorming techniques, teams of students explore and record what could be done with each part to maximize its contribution to overall performance.*

*There are potential safety hazards in the design process that should be discussed at the outset. The hook eyes must be in the body firmly so that the car remains on the track. If you are using handmade wooden blocks, students will need to use hand tools to shape them. Before allowing students to use tools, provide specific safety instructions. If sharp cutting tools are used, you must keep a strict inventory. Set up a hand tool station in full view near your teaching center. Make sure that the bodies of the cars are not so thin or fragile that they come apart during racing.*

*As students discuss the safety constraints of the model cars, extend their understanding to the challenges of real automobile designs. Because the $CO_2$ vehicle moves at a relatively high speed, keeping it on the track is a serious challenge. This unit uses a hook eye that holds the car on a grooved track. On real automobiles human drivers perform that function. In a later activity students also will investigate center of gravity, a factor in vehicle rollover accidents.*

## Conceptual Development

Although design is a highly creative activity, creative freedom is often limited by conditions beyond the designer's control. These conditions are referred to as design constraints. In engineering the constraints are incorporated into the requirements specified by the contractor or by a marketing department on behalf of a consumer. A car's constraints would include safety, speed, appearance, and cost efficiency.

Most students begin by assuming that each kit will result in an identical vehicle. This is partially the result of science lessons that emphasize one right answer. Allocate time for brainstorming. Emphasize that considerable room for creativity exists within the requirements and constraints of this project.

## Extensions

Ask students to imagine that they are engineering a simple vehicle or appliance that will be transported and will operate on Mars. Have them brainstorm the constraints that would restrict the design of something that operated on Mars.

The Technology Student Association (TSA) has formal $CO_2$ vehicle competition standards. For information on this topic, contact TSA directly at 1914 Association Drive, Reston, VA 22091 (phone: 703-860-9000; fax: 703-758-4852).

# Activity 2:
# Mass and Motion

*Question: How does mass affect a car's motion?*

## Objectives

- To define and measure an object's mass

- To discover the relationship between mass and motion

- To measure the motion caused by a constant force acting on different masses

- To graph mass vs. velocity when force is constant

## Materials, per team

- Toy truck

- Four 250-gram masses

- Balance (if masses are not marked)

- 2.5-N spring scale

- Meterstick

- Stopwatch

## Procedure

You probably played with a toy truck as a small child. This time you will explore the truck's motion scientifically. You must control as many variables as possible. One variable is the truck's operator. Begin by having each member of your team try to roll the truck steadily and evenly with a constant force. Decide who can do it most reliably, and elect that person as the operator for the entire experiment.

1. Measure the mass of the truck and record in Data Table One.

2. Hook the 2.5-N scale to the truck. Pull the truck slowly across the table so that the force you apply is 1 N all the way. Remember, inertia makes it harder to start the truck than to keep it going. If the truck's front end tends to lift up from the surface of the table as you pull, you are applying some of the 1-N force upward, so be sure you are pulling horizontally. Every member of your team should practice pulling the truck, in order to feel the effect of a 1-N force. Describe the experience of pulling with a constant force of 1 newton.

   *A force of 1 newton on a mass of _____ is like _____*

   _____

3. You are going to vary the masses on the truck and then measure its speed keeping the force of 1 newton constant. Record the total mass of the truck (including the load you have placed on it) in the first column of Data Table One. For each load measure the total travel time down the track and record it in the second column labeled "1st trial." Conduct three trials for each load.

## Data Table One

| TRUCK'S MASS PLUS LOAD (grams) | TIME TRIALS (SECONDS) | | |
|---|---|---|---|
| | 1st Trial | 2nd Trial | 3rd Trial |
| | | | |
| | | | |
| | | | |
| | | | |

Empty Truck Mass=_____  Force Applied=_____  Distance Pulled=_____

4. Enter the same masses in the first column of Data Table Two. For each load calculate the average time it takes the truck to travel one meter. (Divide the time by the distance in meters.) Record the time in the space provided in the second column of Data Table Two.

## Data Table Two

| TRUCK MASS PLUS LOAD (kg) | AVG. TRAVEL TIME (s) |
|---|---|
| | |
| | |
| | |
| | |

5. Graph the data from Data Table Two, showing the mass of the truck versus the average time to travel the 1-m distance.

# Unit 2: Activity 2

## Mass vs. Time Graph

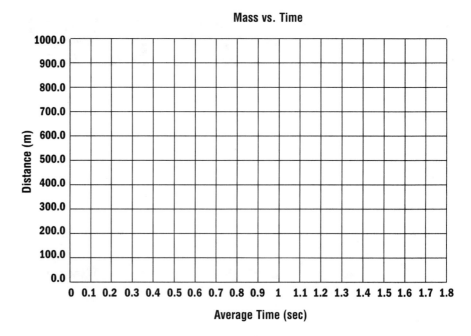

6. Discuss your graph with your team. What is the relationship between mass and time?

_____

Determine whether the relationship is **direct** (force increasing with distance) or **inverse** (force decreasing with distance).

7. What is the meaning of the mass-time relationship for your truck? Discuss this question as a team, and record the discussion in the space below.

_____

_____

_____

_____

_____

_____

NATIONAL SCIENCE TEACHERS ASSOCIATION

8. Look at the list of ideas you recorded in Activity 1 for improving the performance of the various parts of your $CO_2$ car. Considering what you've just learned about mass and motion, how would you change your design ideas now?

_____

_____

_____

_____

_____

9. Enter the definition for mass and other new concepts you have learned into your notebook.

# Teacher's Guide to Activity 2:
# Mass and Motion

*Students explore the relationship between mass and motion by pulling a toy truck with a constant force. In doing so they encounter the effects of inertia. Since the force of the $CO_2$ cartridge will be constant, students can infer that reducing mass will improve the performance of their cars. However, they also can infer that it is difficult to keep a very light vehicle on a steady path, so there will be a limit to the improvement that lightening the car can achieve. There are also safety factors involved in very light or very thin vehicles.*

*Pulling the toy truck with a constant force takes some coordination, an ability to act quickly, and a fair amount of practice. Middle school students enjoy this challenge and can get quite good data, considering the possible errors. Be conscious of gender equity within groups and encourage mutual respect.*

## Conceptual Development

Students may never before have considered the concept of mass as a property that limits motion. This activity may be their first experience in directly exploring the effect of inertia. Try to get students to shift their thinking: Say "our car has mass of 45 g" rather than "our car weighs 45 g." On Earth this distinction is sometimes not so important, but it is essential to space exploration. Also, students should speak in these terms in front of the class during their presentations at the end of this unit.

This activity may well be the first time that students have experienced the effects of inertia in a controlled situation. Most students intuitively think that a constant force keeps an object moving at a constant speed; they do not realize that if the applied force exceeds the opposing frictional force, the object will speed up. If just barely enough force is applied to keep an object moving, students are actually measuring the coefficient of friction ($\mathbf{F_f}$).

Topic: mass

Go to: *www.scilinks.org*

Code: FBP074

Students vary the total mass of the toy truck by adding 250-g masses to it and then measuring the time it takes for the truck to cover a set distance under a constant applied force of one newton. By using a track longer than one meter and then dividing, they minimize the errors due to inertia. They increase the accuracy of their time measurements by running the experiment three times for each load and then averaging the results. They then use the averaged values of the travel time in establishing a relationship between mass and time. Mass is also an important factor in vehicle stability. Students can infer that heavier vehicles hold the road better, and extend this lesson to later discussions on safety.

Students analyze a mathematical function graphically. They determine whether the relationship is **direct** (force increasing with distance) or **inverse** (force decreasing with distance). These terms were introduced in Unit 1 but may require review.

## Extensions

Students may have followed buses or trucks on the interstates and have an intuitive understanding of how long it takes a massive body to accelerate. Ask students to observe the acceleration/deceleration lanes on steep grades on highways and discuss their purpose.

Topic: inertia

Go to: *www.scilinks.org*

Code: FBP075

# Activity 3:
# Force Diagrams

*Question: How can you graphically portray the forces acting on your car?*

## Objectives

- To define and measure an object's weight as a force

- To depict a force as an arrow with magnitude and direction

- To diagram forces acting on an object

- To evaluate the sum of all the forces (the net force)

## Materials, per team

- Toy truck

- 2.5-N spring scale

- Ruler

## Procedure

In Activity 2 you measured the effect of a force on a toy truck with different loads. Now your team will learn how scientists diagram forces. It is important that all your team members understand what each force arrow means. Good team members make sure that everyone is learning the material.

You can answer many of these questions with arrows. Make sure you understand what you are drawing. Each of your team members will be asked to explain them.

1. Weigh your toy truck with the 2.5-N spring scale, and record the measurement in the space provided below. Remember that weight is due to the force of gravity and is not the mass in grams that you measured in Activity 2.

    Weight = _____N

2. Each part of the truck has mass and weight. But we can find one point on the truck where it balances perfectly. This is called the center of gravity. It can be used to represent all of the force on the truck. On the diagram of the toy truck, label where you think the center of gravity is with a large **X**. (You can experiment by try-ing to balance the truck on your hand or a pencil.) Then show the force of gravity by drawing an arrow with its tail at the center of gravity of your car and its head directly downwards. Label this arrow as **Fg** (the g stands for gravity). The magnitude (size) of **Fg** is the weight in newtons measured in step 1. Use a scale 2 cm = 1 N. Divide your truck's weight by two and make your arrow that long in centimeters.

**Force of Gravity**

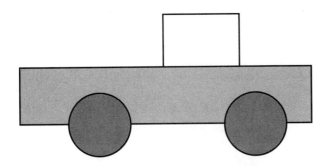

3. When your truck pushes down on the table or floor, the floor pushes back up. If the truck is still, the force pushing up must be exactly the same as the force pushing down. This is called the normal force (**F**$_N$). Draw the normal force by drawing an arrow straight up from the center of gravity on the diagram used in step 2. The upward force is actually exerted on the wheels, but it can be summarized as this one arrow at the center of gravity. Label the arrow as **F**$_N$ and make it just as long as the arrow that depicts the force of gravity.

4. The diagram is now a force diagram of the toy truck when it is standing still. There is no unbalanced force. We use the term net force (**F**$_{net}$) to indicate the difference between the forces acting on an object. If there is a net force, an object moves or stops. But your truck is standing still. To indicate that there is no un-balanced force, write **F**$_{net}$ = 0 above the diagram. You have now made a complete force diagram.

5. Make a force diagram of the toy truck as it is being pulled by a 1-N force, as in Activity 2. On the next diagram, label the center of gravity with a large **X**. Next, draw an arrow representing the 1-N pulling force, and label this new arrow as **F**$_{applied}$. Use the same scale as indicated in step 2: 2 cm = 1 N.

**Pulling Force**

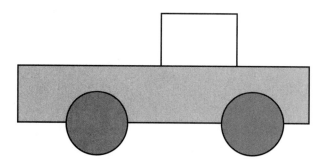

6. Remember that the force of gravity, **F$_g$**, and the normal force, **F$_N$**, are still acting. But the truck is not going up or down so they are still equal. Draw and label these forces on the figure. Make them the same length, using the scale 2 cm = 1 N. Net force is zero.

7. There is a force holding the truck back. It is friction (**F$_f$**). Make an arrow for the force of friction. Estimate its length and label it **F$_f$**. You will measure the exact magnitude of this force in Activity 4. Then make an arrow for the net force moving the truck forward and label it **F$_{net}$**. When the truck just begins moving, the net force must be greater than the force of friction but less than the force on your spring scale. Even though you don't know its exact magnitude, you should be able to make a good estimate from your experience.

8. Check your team members' force diagrams, and discuss any differences. Now discuss the ways in which the toy truck is similar to your $CO_2$ car. How can force diagrams and the concept of the center of gravity be used in designing $CO_2$ cars? Look at the list of design ideas from Activities 1 and 2. Should the applied force from the $CO_2$ cartridge go through the center of gravity of your car? Record your responses and ideas in the space below.

_____

_____

_____

9. Remember to record the definitions of new concepts and processes in your notebook.

# Teacher's Guide to Activity 3: Force Diagrams

*Students are informed that force diagrams are a graphic way of describing the forces acting on their toy truck in Activity 2. They start by measuring the force of gravity exerted on the toy truck, with a reminder that forces of gravity (or weight) and mass are different. Students are then guided through the process of drawing force diagrams, first of a toy truck standing still and then of a toy truck just starting to be pulled, as in Activity 2.*

## Conceptual Development

This activity may stimulate a discussion of the use of the words "mass" and "weight," as well as the units in which they are expressed. This discussion should establish a scientific definition of weight as the result of the force of gravity, which is measured as a scale reading in units of newtons.

Students also begin exploring forces that act in opposite directions. They infer that if the net force is not equal to zero, motion or change in motion will occur. In a truck that remains horizontal on the track, the force of gravity and normal force are equal and no net motion occurs. Students diagram the opposing forces of pull (applied force) and friction and infer that the force applied must be greater than the force of friction for the truck to begin to move forward. In this exercise students estimate a magnitude for friction. In the next exercise they measure it quantitatively. The estimation process provides a good opportunity to examine students' preconceptions about these forces.

The most important ideas that students need to develop are notions that forces can be depicted graphically as arrows and that net forces result in changes in motion. The process of drawing force diagrams is an example of scientific modeling. Remind students that these diagrams are a simplification of reality. Professional engineers would examine the force on each part of a vehicle in the design process.

There are two subtle ideas that emerge: net force and center of gravity. Students should realize that the motion due to two or more

Topic: forces

Go to: *www.scilinks.org*

Code: FBP079

forces acting on an object simultaneously is equivalent to the motion that would occur if just a single force (namely, the net force) were acting. In addition, the point at which a force acts is important. For a force to cause an object to move in a straight line, the net force must act through the center of gravity in the direction of that line.

The directions take some liberty with the idea of normal forces. In fact, there should be two arrows representing the normal force on the tires, with their total magnitude equal to the force of gravity so that net force is equal to zero. Because the principal concept of this lesson is center of gravity and because students are only beginning to understand summing forces that act in opposite directions, this idea has been simplified.

This exercise considers only forces that act in opposite directions to one another. To mathematically calculate the resultant (net) force from those that act at angles to one another, trigonometry is required. However, the use of vector diagrams can still provide students with rough estimates of those forces. Following are sample responses for the force diagrams explained in Activity 3.

### Force of Gravity plus Normal Force

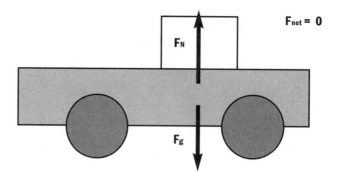

## Applied Force Diagram

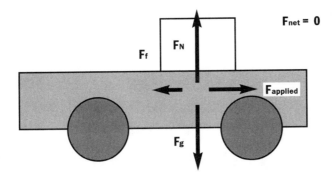

## Extensions

There are several virtual sports computer programs that analyze the forces exerted by the human body on objects such as golf clubs or baseball bats. Many computer games reveal the effects of forces without explanation. Encourage students to connect this activity to programs that use sensors and computer graphics to generate diagrams similar to the ones they have drawn.

Center of gravity is an important concept in the stability of automobiles. It has come under special scrutiny in the rollover rates of certain sport utility vehicles (SUVs). Some critics contend that SUVs have centers of gravity too high for their relatively narrow wheelbases; others argue that SUV drivers tend to overestimate the ability of a four-wheel drive vehicle and drive more carelessly. Collect very clear pictures of various vehicles including sports cars, sedans, and SUVs, and have students mark the center of gravity for a bulletin board display on stability.

The nation's largest learning laboratory on transportation safety is the Crash Injury Research and Engineering Network (CIREN). Force diagrams of actual crashes are available from CIREN through the National Highway Traffic Safety Association (NHTSA) webpage: *www.nhtsa.dot.gov.*

# Activity 4:
# Frictional Force

## Question: What is friction and how does it affect motion?

## Objectives

- To define and measure friction as a force

- To draw force diagrams that portray significant frictional forces

## Materials, per team

- Toy truck

- 2.5-N spring scale

- Several 2.45-N weights

- Ruler

## Procedure

In Activity 3 your team learned how to draw a force diagram for a toy truck that was just starting to speed up. In this activity you will measure the magnitude of the frictional force on a truck with a heavy load going with steady velocity.

1. Friction is the force that resists motion of one material over or through another. A toy truck has many parts that must move over others while the truck runs. Make a list of all the places friction could occur on the truck.

_____

_____

_____

2. You know that it is harder to start the truck than to keep it going. Starting the truck requires you to overcome inertia. After the truck starts moving, keeping it going at a constant speed requires you to overcome friction. Hook the 2.5-N spring scale to your truck. Add two 2.45-N masses to the load. Practice pulling the truck several times, so that you can keep it going at a constant speed. Record both the force necessary to keep the truck moving after it gets started and the total weight of the truck and load:

$$\mathbf{F_f} = \underline{\hspace{2cm}} \text{for a weight of} \underline{\hspace{2.5cm}} N$$

3. On the Net Force Diagram below, draw a force diagram of the toy truck moving at a constant speed. Because the motion is constant, $F_{applied}$, is equal to the frictional force, $F_f$, measured in step 2. Follow the steps below, and use the scale 2 cm = 1 N.

- Place an **X** to indicate the center of gravity.
- Draw an appropriate-length arrow for the force of gravity (weight), and label it $F_g$.
- Draw an appropriate-length arrow for the upward (normal) force. Label it $F_N$. Remember, if the truck is on a level surface $F_N$ is equal to $F_g$.
- Think of your truck as moving steadily forward in a straight line. Draw $F_f$ and $F_{applied}$ in opposite directions.
- Indicate the net forces vertically and horizontally. Label these $F_{net}$. If the net forces are zero, indicate this by writing the equation $F_{net} = 0$ above the truck.

**Net Force Diagram**

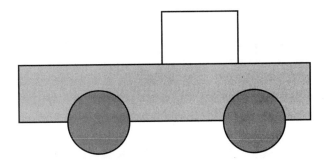

4. Imagine that you could apply a steady force of 5 N. Friction would stay the same so there would be a net force acting. Draw the force diagram. Use the scale 2 cm = 1 N.

**Applied Force Diagram**

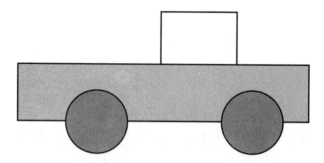

What will happen to the truck's motion?

_____

_____

5. Now imagine that you load the truck with the same weights and apply a 10-N force. Draw a force diagram of this situation. Use the scale 2 cm = 1 N.

**Larger Applied Force Diagram**

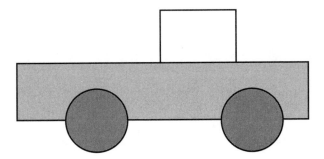

6. Discuss this material as a team, to be sure that you all understand the three force diagrams. Discuss how these diagrams are related to the design of your $CO_2$ car. Which arrow represents the force of the $CO_2$? How can you make $F_f$ less? Record your team's ideas in the space below. Use arrows and diagrams.

7. Review your team's $CO_2$ car design ideas from Activity 2. Record how your team's design incorporates what you've learned.

# Teacher's Guide to Activity 4:
# Frictional Force

*Using the toy trucks, students measure frictional forces and make corresponding force diagrams. They then imagine two situations in which the trucks are accelerated as a result of even larger applied forces. Students practice drawing these force diagrams and relate the diagrams to their $CO_2$ car designs. Students need to understand that frictional forces partially cancel the applied force.*

## Conceptual Development

While reinforcing their new knowledge of force diagrams, students measure friction as a force that opposes motion. To measure friction, students must develop an understanding of the concept of inertia: A body at rest remains at rest and a body in motion remains in motion unless acted upon by an unbalanced force.

Students learn that friction is a force that opposes the motion of a body over or through another, measured in newtons (N). The coefficient of friction for two sliding materials is obtained by dividing the force of friction by the force of gravity. For example, the coefficient of friction of dry oak on dry oak is approximately 0.3; the force of friction would be approximately 0.3 x the weight of the sliding block. For rubber on dry asphalt it is 1.07, but when the pavement is wet it is reduced to 0.95. Students may have the misconception that low friction would make better tires. They should realize that the friction of the rubber on the pavement allows the car to accelerate. It is the friction within the vehicle on its moving parts that must be reduced.

Students should come to understand that friction can be overcome by other forces, but that mass is an intrinsic property of an object. Mass (weight) increases the effect of friction because it increases the contacts among moving parts. Students also should learn that if the force that is applied is greater than frictional forces, there will be a change in speed. (They may not yet understand the term "acceleration.") If you have access to larger capacity metric spring scales, you can have students physically explore the effect of increased force.

Topic: friction

Go to: *www.scilinks.org*

Code: FBP087

The idea that these forces act on the center of gravity of an object should prompt students to think about designing their cars in a balanced way, so that the center of gravity is in front of the $CO_2$ cartridge. They should also begin to understand that to increase the net forward force on their car (with $CO_2$ constant) they must reduce friction. Look for these ideas in the response to question 6. Below are sample responses for force diagrams explained in Activity 4.

### Net Force Diagram Answer

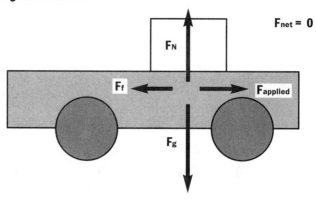

### Applied Force Diagram Answer

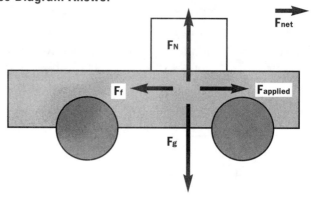

### Larger Applied Force Diagram Answer

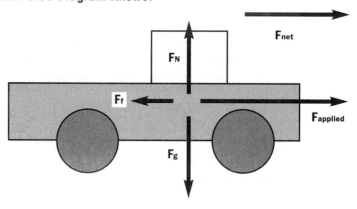

## Extensions

When Isaac Newton first proposed the law of inertia, the idea of a frictionless environment was only theoretical. But today's students can examine the trajectory of space probes in the nearly frictionless environment of space. Have students access NASA's website *www.NASA.gov* to read reports on the progress of Voyager 1 and 2. That will allow them to view an object "in motion which remains in motion." They also can examine the effects of the forces that still act on these space probes, such as the gravity of Jupiter and Saturn, to change the direction of the probes' motion.

Auto tires are designed to increase the ability of the car to hug the road while reducing friction. Have students access gliding friction. Using a digital camera, take close-up photos of automobile tire treads and have students classify them as high-friction and low-friction designs. High-friction tires enable cars to accelerate more quickly, but these tires result in greater consumption of fuel.

Topic: Sir Isaac Newton

Go to: *www.scilinks.org*

Code: FBP089

# Activity 5:
# Writing Design Briefs

*Questions: How can an engineering team explore many ideas for a prototype quickly?*

## Objectives

- To explore ideas using thumbnail sketches

- To design a prototype within constraints

- To develop specifications through sketching

## Materials, per team

- Copy of *Sketching Guide*

- No. 2 pencil and sketching paper

- Grid paper for making final drawing

## Procedure

In this activity you will practice communicating ideas quickly and efficiently with members of your team. To do this you will practice sketching and drawing.

1. In Activity 1 you brainstormed with your team concerning your goals (product requirements) for a $CO_2$ vehicle and then recorded them in your notebook. Discuss these goals again with your team. When you have decided on the requirements for your vehicle, list them here:

_____

_____

_____

_____

_____

_____

_____

_____

2. Make your first thumbnail sketch of how you think your car should look. Remember to add the source of force, the $CO_2$ cartridge. Put this in your notebook.

3. Next we will use a technique that engineers call iteration to improve the design. Draw your car again and again. Each time you do, make one improvement based on what you know about force, center of gravity, and friction. You can read about iteration in the *Sketching Guide*.

4. After you have explored many ideas with sketches, show your ideas to your team and look at those of the other members. Select the idea that you believe will best meet the design specifications and goals.

5. Decide on the dimensions of your car. If your kit includes rules for placing axles or screws, make sure you follow them. Decide on a scale for your drawing. (For example, you might decide that the car will be 15 cm long. You might decide that each block of the grid paper will represent 1 cm.) Make sure that you have enough space on the paper to fit the entire car and the labels.

6. Using graph paper, make side-view and top-view drawings of your design. On your drawings, be sure that you place the hole for the $CO_2$ cartridge in an appropriate spot.

# Teacher's Guide to Activity 5:
# Writing Design Briefs

*To work out details of the vehicle design, students use pencils to sketch ideas on both sketching paper and graph paper. They also will be introduced to iteration techniques as a tool for working out design ideas. The ideas should reflect the criteria for maximizing the design within the constraints identified earlier.*

## Conceptual Development

Students learn the importance of working within limitations or constraints. By making many quick drawings, students will learn that, even with constraints, many creative possibilities still exist.

Scaling a drawing may be new to students. They may need extra sheets of paper because they will make errors. Encourage them to use a very simple scale. If possible, use paper with 1-cm grids so that students can draw to actual size. If you opt to have students create their own car bodies rather than use purchased kits, the scaled drawings will be the actual plans for shaping the wooden chassis. Encourage them to think through all of the dimensions (including wheels and the size of the cartridge) before they finalize their drawing of the car body.

## Extensions

The criteria for vehicle performance are frequently in conflict with the criteria for vehicle style. Students may collect photographs of automobile styles and classify them according to vehicle safety ratings (available in the media or on the Web).

# Activity 6:
# Prototype Development

*Question: Which steps are needed to develop a prototype?*
*How can the prototype be manufactured accurately and efficiently?*

## Procedure

Before you begin working on the construction of your $CO_2$ car, you must write a manufacturing plan. This is a list of specifications and the steps you must follow to create your vehicle.

1. Your instructor will furnish you with a form entitled "Prototype Manufacturing Plan." List all the things you have to do in order to build your car. Identify which team member will be responsible for each step. Your instructor must sign the plan sheet before your team begins work on your car. Keep the plan in your notebook for future reference.

**NOTE: The following steps should be divided among team members.**

2. Using your plans for the shape of your team's car, cut out the patterns and tape them to the vehicle block to decide which works best. Write this step in the appropriate place on your "Prototype Manufacturing Plan" handout.

3. Review the safety rules for cutting with your team.

4. When your team is prepared to follow safety rules, one student should move to the cutting station and carefully shape the chassis of your team's vehicle to match the plan you have developed.

5. Use sandpaper to finish shaping the team vehicle.

6. Attach the wheels to the car. Attach the screw or hook that will keep the car on the track. Prepare the opening to insert the $CO_2$ cartridge and test that it fits.

## Objectives
- To create a manufacturing plan that will result in a prototype that corresponds to the design goal
- To carry out a manufacturing plan in a manner that ensures maximum quality control

## Materials
- "Prototype Manufacturing Plan" handout for each student
- Fabrication or prototyping lab equipped with the following tools: handsaw, hand drill, sandpaper, files, cutting tools with safety guards
- Finishing material (acrylic paint, glues)
- Testing ramp (plywood or access to the school's handicap ramp)

7. Test to see if the wheels are aligned properly. Perform a gravity roll for the vehicle (down an inclined ramp) to determine wheel alignment accuracy and friction drag. Adjust as needed.

8. When the team car is completely shaped, you may paint it. However, you may choose to test the car in Activity 7 before you put the final paint on the car. Follow each step of your manufacturing plan until the car body is complete.

# Prototype Manufacturing Plan

Name _____

Period____ Grade ___ Notebook No. ___

Start Date _____

Complete Date _____

Start Approval Date          Finish Approval Date

Prototype of _____

## Thumbnail Sketch of Project
(Staple full technical sketch to back of Prototype Manufacturing Plan)

Procedures                                        Tools and Equipment

                                                 Material Used

Your manufacturing plan must be approved before you begin prototype construction and after the prototype is complete.

# Teacher's Guide to Activity 6:
# Prototype Development

*Students create a manufacturing plan for the prototype and then carry out the plan in a safe manner in the manufacturing lab. Students identify the tools they need to carry out each step of the plan. Students then construct the prototype according to their manufacturing plan, divide and delegate responsibility, and follow safety rules.*

## Conceptual Development

To develop a quality product, students must apply the principles they have learned. Planning, delegation of responsibility, and time management are all part of the manufacturing plan. Students develop a degree of skill and understanding of the interaction of tools and materials in the process of parts fabrication.

Students will be anxious to begin this activity. The most important part of classroom management may be encouraging deliberation and caution. Remind them that if they cut away part of a wooden block, they cannot replace material that is removed.

Reinforce the connection between the scientific investigations and the manufacturing plan by asking good questions: "Where is the center of gravity of that vehicle?" "Is the $CO_2$ cartridge going to be aligned with the center of gravity?"

Safety is of primary importance in classroom management. Set up a specific station for the use of a handsaw or a hand drill within full view of the teaching station. Limit use to one student at a time. Purchase cutting blades with safety shields, and inventory them carefully at the end of class. Tolerate no horseplay or clowning around in the manufacturing process.

If you use paints or glues, you must make sure that you have a Materials Safety Data Sheet on hand for each product in the classroom. Ensure that there is adequate ventilation. You may need to restrict painting to one station so that odors and fumes do not rise to a level that could bother sensitive students. Do not bring flammable solvents into the classroom.

Remember that you are responsible for any safety hazards that could result from homework assignments that you make. Therefore it is usually better to have all construction occur in your classroom rather than to delegate supervision to others.

## Extensions

The Occupational Health and Safety Administration (OSHA) regulates the safety procedures in manufacturing facilities. Since students are modeling a real manufacturing process, they should investigate and adhere to all OSHA requirements. In most laboratory situations OSHA recommends a ventilation system capable of changing the air in the room six times per hour.

# Activity 7:
# Performance Testing

*Question: How can performance be maximized in a $CO_2$ car?*

## Objectives

- To develop a testing procedure for performance

- To carry out testing procedures and record data

- To evaluate test results

- To suggest design modifications incorporating the best ideas

## Materials

- Triple-beam balance or electronic scale

- $CO_2$ car specifications, per team

- Stopwatch, per team

- $CO_2$ car test track

- Spreadsheet for recording results, per team

- Material or substance for reducing friction (lubricant such as graphite), per team

## Procedure

1. Weigh your finished vehicle and place its weight on a spreadsheet of class results.

2. Carefully examine each of the vehicles designed by students in the class. Using the criteria for the design, predict which vehicles will have the best performance. Record this prediction here:

_____

_____

_____

3. Prepare your vehicle for testing by reducing friction. You may choose to use graphite as a lubricant. Decide with your team how much to use and where to apply it.

4. Participate in the performance test with your vehicle. Place a $CO_2$ cartridge in the vehicle. When it is your turn, place the vehicle on the track. Appoint one student as timer. Post your results when your test is concluded.

5. Compare the results for your vehicle with those of other teams in the class.

6. Discuss with your team how your vehicle can be improved. Make a list of improvements here.

_____

_____

_____

# Teacher's Guide to Activity 7:
# Performance Testing

*Students examine each of the vehicles and determine the different ways in which each of the design criteria were met. They construct a chart or spreadsheet listing the mass of each vehicle and the approach taken to its design. Students then test the cars and record results. They examine the results to verify predictions and identify which criteria had the greatest effect on vehicle performance. Students then make suggestions for redesign of the vehicles to enhance performance.*

## Conceptual Development

Students learn to predict performance based on the degree to which a vehicle design deals successfully with the design constraints. They also learn to compare and contrast different design solutions and to determine which factors contributed to the success (or lack thereof) for a variety of vehicles. They learn to incorporate test information into design revisions.

An important part of this process is the development of hypotheses. Students should be given ample time to analyze one another's car designs and masses. They should be encouraged to verbalize not only which car they believe will be fastest (which may be influenced by appearance) but also why they believe their predictions will be true. They may relate size, mass, aerodynamic shape, or size of wheels to performance.

## Extensions

Students will have an intuitive understanding of aerodynamic shape and the effect of wind (fluid) friction on performance. Although these effects are very small on cars of this mass and size, they can be quantified. Wind tunnels are available from the manufacturers of car kits for wind resistance testing.

# Activity 8:
# Time and Velocity

*Question: What is the average velocity of a CO₂ car?*

## Objectives

- To measure the time it takes to traverse a set track
- To calculate average velocity
- To describe how a $CO_2$ car's mass, design, and frictional forces affect velocity

## Materials

- Finished $CO_2$ car, per team
- Track
- Stopwatch, per team
- Lubricant, per team

## Procedure

In this activity you will measure the velocity of your $CO_2$ car. Remember, velocity is speed in a specific direction. Since cars are moving away from their starting point, the velocity is a positive number, equal to distance divided by time.

Your team will need to establish who will start your car, who will pick up your car at the end of the track, and who will record the travel times of your car. The velocity of your vehicle will not be the same at every point, so you will need to compute the average velocity over the time of the runs.

1. Prepare your $CO_2$ car for racing. You may wish to adjust the wheels, replace lubricant, or test the placement of the cartridge again.

2. Do two runs of your car, carefully recording the time. Divide the time by the distance of the track to determine the velocity. Then average two runs to determine the average velocity.

   What is the average (mean) velocity of your car? _____

   What is the average (mean) velocity of all the cars?_____

   What is the middle velocity of the cars in the class (the median)? _____

   What is the most common velocity of the cars in the class (the mode)? _____

**Vehicle Mass Data Table**

| TEAM NUMBER | MASS (kg) | TIME (s) | | |
| --- | --- | --- | --- | --- |
| | | 1st Race | 2nd Race | 3rd Race |
| 1 | | | | |
| 2 | | | | |
| 3 | | | | |
| 4 | | | | |
| 5 | | | | |
| 6 | | | | |
| 7 | | | | |
| 8 | | | | |
| 9 | | | | |
| 10 | | | | |
| 11 | | | | |
| 12 | | | | |

3. Can you see any relationships between velocities and other characteristics of the cars? Do the lighter cars go faster? Do more streamlined cars go faster?

_____

_____

_____

4. Convert your car's average velocity from meters per second to kilometers per hour, using the conversion factor 1 m/s = 3.6 km/hr. Does your car go as fast as a real car?

_____

_____

_____

# Unit 2: Activity 8

5. Is the average velocity the same as the $CO_2$ car's top velocity? If not, what do you think the $CO_2$ car's top velocity was? Why?

_____

_____

_____

6. If you were going to build another car, what would you change?

_____

_____

_____

7. Were some of the cars unable to travel straight down a ramp or on the track? These models would be unsafe. If a manufacturer wanted to build a safer car, what changes would it have to make?

_____

_____

_____

# Teacher's Guide to Activity 8:
# Time and Velocity

*During the Great Race students compare the characteristics of cars to their performance velocity. They now have an interesting set of data to explore. Students then relate their discoveries to performance in real automobiles by converting velocities to km/hr.*

*Students develop inferences about the differences in three measures of central tendency: mean, median, and mode. Using a spreadsheet, they can examine the relationship between performance and other parameters of design.*

## Conceptual Development

This exercise allows students to examine group data and make inferences about relationships. Although a chalk board or chart board display can serve as a spreadsheet, it is more efficient to use a computer spreadsheet projected for the class. This technology allows for faster analysis of the data. It can also allow students to generate mathematical questions such as "Do heavier cars go faster?" and rank the data in order to provide rapid feedback.

Students discuss the idea of average velocity in their analyses. They can observe that the velocity of the vehicle increases at first and then decreases. The meaning of average velocity is explored in the contexts of vehicle design and safety.

When students discuss the differences in race times, they start talking in terms of one or two thousandths of a second, as if that were a perfectly normal thing to do. This experience builds understanding of these very small intervals of time and adds interest and meaning to the data.

## Extensions

Students will be reluctant to stop the process of improving their cars' performances. They may suggest different lubricants, wheel alignments, or the addition of supplementary masses. Real engineering efforts work within time constraints; these limitations are an important part of the unit's theme.

Topic: velocity

Go to: *www.scilinks.org*

Code: FBP103

One real-world analogy to these design modifications is the recall process. While most recalls are related to safety issues, occasionally a recall will be announced to improve performance. Students may be able to report examples of product recalls.

All three *Fender Bender* units have connections to vehicle safety. In this unit the relationships among mass, aerodynamics, and velocity are clear. Students may be tempted to make their cars as light as possible to improve performance. Experimentation will reveal that there is a limit to this; beyond that limit the vehicles lose stability. Students will also discover the importance of wheel alignment to performance.

# Teacher's Guide to Unit 3:
## The Space-Frame Vehicle

*This unit explores how forces affect the structure of a vehicle frame. Students design a space frame that will be attached to a wheel block and crashed. Strapped into a vehicle is a proxy passenger—an egg in student tests. The goal for a student design team is to direct forces around the vehicles rather than through them, thus protecting their passenger from injury.*

## Overview

As in the first two *Fender Bender* units, students conduct empirical investigations to provide data for their designs. They begin by exploring how passenger safety can be engineered in a vehicle. Next students construct force-path diagrams that describe crashes. With these graphic models, students see that energy-absorbing structures in vehicles can protect passengers. They then investigate the amount of work it takes to roll their space-frame vehicles to the top of the ramp, and they use their results to predict the stopping force on the vehicle during a crash. They compare materials as they plan their designs.

Students design and build their vehicles from wood. The three sections of the space frame are the front end, passenger compartment, and luggage compartment, and all have size criteria. The style and size of the passenger seat and seatbelt are identical in all the student vehicles. Because the passenger seat is attached to a solid wood base, the vehicle must absorb the force or the egg will crack. Impact-absorbing material can be added to protect the passenger. The final testing is both spectacular and fun.

Students record and analyze the collisions with the help of video technology. They analyze the force paths and energy transfers of the destroyed space frame by observing the performance of the vehicle and the breakage of the egg. Then they document their design process and present their design portfolio to the class.

## Materials for Space-Frame Vehicle Unit

- Copy of student handout, "The Space-Frame Vehicle Challenge," for each student

- Copies of each of the eight activities, along with *Work as Mechanical Energy*, *Student Safety Guide*, and *Sketching Guide*, for each student

- Computers for research

- One 30-cm ruler per team

- Triple-beam balance scale

- Hobby glue*

- Wood glue*

- One stick of balsa—3 mm square and 30 cm long—per team

- Four sticks of balsa—3 mm square and 6 cm long—per team

- One birch dowel rod—3 mm square and 30 cm long—per team

- One plastic rod—3 mm square and 30 cm long—per team

- One small bucket with handle (child's sand bucket) per team

- About 0.1 kg sand per team

*continued on next page*

### Materials *continued*

- One utility knife with protective guard per team

- Handsaws (miter box optional)

- Energy-absorbing materials such as foam for each team

- Egg (passenger) compartment made from egg carton

- 1-kg mass with hook per team

- 5-kg mass with hook per team

- Eight plastic tie straps per team

- One 2.5-N spring scale per team

- One meterstick per team

- One large testing ramp (plywood or ramp outside of school)

- One large egg per team, plus some extras

- One wheel-block assembly per team

- Eye protection

If you did not cover the mousetrap car unit, you will need to adapt "Activity 1: Writing Design Briefs" and "Activity 3: Writing Design Process Papers" for use in this unit. You also may wish to use force diagrams—covered in Activity 3 of the $CO_2$ car unit—as a review.

## Conceptual Development

To launch the unit, introduce the challenge of designing and building the car by distributing the student handout, "The Space-Frame Vehicle Challenge." The handout gives students an overview of expectations and a table for recording points earned. Tell students to make the handout the first page of their notebooks.

Students develop their understanding of forces and motion intuitively as small children. But many of their naïve ideas prove to be incorrect. By using diagrams to describe forces, students can communicate ideas and improve understanding. This graphic modeling technique is more intuitive than quantitative, but students often consider it abstract when they first begin. The proof of ideas about forces comes in the destructive testing of the space frames.

Students apply a force over a distance in order to pull a vehicle up a ramp. They learn that they have done work on a vehicle and increased its potential energy. They apply this information to an analysis of the forces that will stop the vehicle. This sequence of concepts will be new and difficult to most middle school students.

In addition to examining forces in new ways, students develop spatial perception as they construct various scaled technical drawings and then translate them to the full-size, three-dimensional prototype. Students learn that vehicles can be designed to withstand large forces. They also explore various safety features of vehicles and roadways that minimize the effects of large, potentially hazardous stopping forces. For an overview of the eight instructional activities and how they link with the design process, see the diagram on the next page.

## Space-Frame Vehicle Challenge
### Design Process With Associated Instructional Activities

```
                    ┌─────────────────────────────┐
                    │     DEFINE THE PROBLEM      │
                    │  Activity 1: Passenger Safety │
                    └─────────────────────────────┘

┌──────────────────────────────┐      ┌──────────────────────────────────┐
│      GATHER INFORMATION      │      │          EXPLORE IDEAS           │
│ Activity 2: Force-Path Diagrams I │      │ Activity 4: Testing Properties of Materials │
│ Activity 3: Force-Path Diagrams II │      └──────────────────────────────────┘
└──────────────────────────────┘

                    ┌─────────────────────────────┐
                    │     DEVELOP THE DESIGN      │
                    │ Activity 5: Design Documentation │
                    └─────────────────────────────┘

┌──────────────────────────────┐      ┌──────────────────────────────────┐
│     CONSTRUCT PROTOTYPE      │      │        TEST AND EVALUATE         │
│ Activity 5: Design Documentation │      │ Activity 6: Work and Energy      │
└──────────────────────────────┘      └──────────────────────────────────┘

                    ┌─────────────────────────────────────┐
                    │              REDESIGN               │
                    │ Activity 7: Destructive Testing and Analysis │
                    │ Activity 8: Energy Transfers        │
                    └─────────────────────────────────────┘
```

**Team Project Presentation**

## Tips for Making the Unit Work

*Arrange the classroom for movement and active learning.* You will need adequate flat table space. Design and construction involves close-up, detailed teamwork. The destructive testing is best done outside.

*Emphasize safety.* Cutting and gluing activities involve certain hazards. Cutting the the balsa wood could damage tabletops. You might want to establish a single cutting station or provide cutting boards or old cafeteria trays. Trays or paper can protect tabletops from glue. Maintain Material Safety Data Sheets on all glue products. Insist on eye protection when cutting.

Interact with each student frequently. The nature of the activities in this unit provides plenty of opportunity to work with each team. Show students individually how to make good balsa-wood joints by cutting the angles properly, applying the right amount of glue, and pinning them to dry.

*Communicate high expectations.* Most students have never built a structure of this type, but it is a task almost all enjoy and perform quite well.

*Give praise and encouragement.* If students have difficulty coming up with their own designs, encourage them to use one of the side or top views suggested in Activities 2 and 3. Students can be creative in the ways in which they connect them to make a 3-D structure.

*Communicate affection and regard.* Enjoy the process. Things will get serious later when students' passenger eggs are endangered.

*Give students responsibility.* Students will bring friends and parents in to see their vehicles since they won't last to take home. Ask students to help with organization and display.

*Treat all students equally and fairly.* Make sure the wheel blocks, eggs, and runs are identical. Insure that each team member has an equal opportunity to participate in all activities. Girls and boys can enjoy equal success.

## Physics Background

Encourage students to create a glossary section in their notebooks for definitions and symbols used in their force and motion investigations, such as those listed below, rather than scattering definitions and symbols in many locations.

**Work (W):** a force acting over a distance ($W = F \times d$), the transfer of energy, measured in joules (J). One joule is the force of one newton (**N**) acting over a meter (m).

**Energy (E):** the ability to do work, measured in joules (J)

**Potential Energy:** stored energy, measured in joules

**Kinetic Energy:** energy of motion, measured in joules

**Force Path:** the point at which a force acts on an object, together with the direction of that force. On a force-path diagram each force is designated by an arrow that shows its point of application and its direction.

## Time Management

A suggested schedule for covering all the activities of this unit over the course of 20 school days (four five-day weeks) is given in the table below. It is based on allotting one class period of 45 minutes per day to the unit.

### Time Management for Space-Frame Unit

| DAY NUMBER | ACTIVITY |
|---|---|
| 1 | Space-Frame Vehicle Challenge Unit Overview |
| 2 | Activity 1: Passenger Safety |
| 3 | Activity 2: Force-Path Diagrams I |
| 4 | Finish Activity 2: Force-Path Diagrams I |
| 5 | Activity 3: Force-Path Diagrams II |
| 6 | Finish Activity 3: Force-Path Diagrams II |
| 7 | Activity 4: Testing Material Properties |
| 8 | Finish Activity 4: Testing Material Properties |
| 9 | Presentation Templates; Activity 5: Design Documentation |
| 10 | Finish Activity 5: Design Documentation |
| 11 | Build Space-Frame Sides (2-D) |
| 12 | Build Space-Frame Sides (2-D) |
| 13 | Build Space-Frame Sides (2-D) |
| 14 | Activity 6: Work and Energy |
| 15 | Finish Activity 6: Work and Energy |
| 16 | Activity 7: Destructive Testing and Analysis |
| 17 | Activity 8: Energy Transfers |
| 18 | Portfolio and Presentation; Preparation |
| 19 | Presentations; Portfolios Due |
| 20 | Presentations; Portfolios Due |

## Unit Assessment Criteria

The "Space-Frame Vehicle" unit affords students the opportunity to intuitively design a structure. There are no right answers. Students should receive a good deal of credit for this process. The destructive testing is quick and somewhat unpredictable, so the bonus points granted for any passenger eggs that survive are well deserved. You can assess concept development by periodically asking questions that can be answered by force diagrams. Questioning team members about their work can help ensure that each member of the team is participating fully.

The goals of the "Space-Frame Vehicle" unit include organization, creativity, scientific inquiry, and concept development. To accomplish those goals, embedded assessments are more appropriate than end-of-unit tests.

## Notebook

If students have completed Units 1 and 2, they should be proficient at creating an organized notebook. Assess their organization with periodic questions about specific contents.

## Design Portfolio and Oral Presentation

You may wish to add other types of assessment to the evaluation system suggested here. This scheme places a priority on technical communications, including the design portfolios and oral presentations. Discuss your priorities in advance with the students so that they can become responsible for their own success. Points for the oral presentation might be discussed when you distribute the student handout, "The Space-Frame Vehicle Challenge."

# Introduction:
# The Space-Frame Vehicle Challenge

*Question: What is the problem, and what function(s) does the design need to accomplish?*

Your challenge is to design and build a space-frame vehicle—using only the materials supplied—that will prevent an egg from being broken during a front-end collision. The egg is supposed to represent a passenger in your vehicle.

You will be part of a Vehicle Research Team (VRT) that will investigate the forces that affect a vehicle in a crash. You will test a variety of materials and select those that you believe are most likely to protect the passenger. Then you will build your vehicle and test it with your passenger aboard.

Organization is an important factor in engineering. You will be responsible for a notebook that documents what you learn and each step in your design. At the end of the unit you will prepare a design portfolio containing your test data and design along with a written summary of your design process. Your VRT will make a two-to five-minute presentation to the class that includes the following information:

- The goals and requirements of your project
- The testing data that informed your design
- How you designed your vehicle to protect the passenger
- How your design performed in testing
- How your design could be improved
- How the principles of these activities are applied to car design and protective gear

Because the process is the most important part of the project, you will be graded on your notebook, your portfolio, and your presentation as well as your activities. The chart on the next page, to be filled in throughout the unit, shows the products and activities that will make up your project.

# Unit 3: Introduction

## Unit Point Values and Grade Scale

| PORTFOLIO TABLE OF CONTENTS | DATE ASSIGNED | PERCENT OF GRADE | POINTS EARNED |
|---|---|---|---|
| 1  Unit Student Handout (this paper) | | | |
| 2  Design Portfolio Technical Paper | | | |
| 3  Assembled Portfolio | | | |
| 4  Team Project Presentation | | | |
| 5  Activity 1: Passenger Safety | | | |
| 6  Activity 2: Force-Path Diagrams I | | | |
| 7  Activity 3: Force-Path Diagrams II | | | |
| 8  Activity 4: Testing Material Properties | | | |
| 9  Activity 5: Design Documentation | | | |
| 10  Activity 6: Work and Energy | | | |
| 11  Activity 7: Destructive Testing and Analysis | | | |
| 12  Activity 8: Energy Transfers | | | |
| 13  Notebook | | | |
| TOTAL | | | |

Performance Bonus (extra credit)

Grade Scale

GRADE

# Activity 1:
# Passenger Safety

*Question: How can a space-frame vehicle be designed and built to maximize passenger (egg) safety?*

## Procedure

In this unit your ultimate task will be to design a space-frame vehicle that will optimize passenger (egg) safety. A space frame (also known as a chassis) is a structure that surrounds the passenger space, engine, and luggage compartment. The vehicle you design will crash, but hopefully your passenger will survive.

1. View the websites listed below and check out the crash tests. What kind of damage is most common in vehicle crashes?

_____

_____

- *www.nhtsa.dot.gov*          - *www.cic.cranfield.ac.uk*
- *www.nhtsa.dot.gov/kids*      - *www.crashtest.com*
- *www.nrma.com.au*

2. There are three main compartments in a passenger vehicle: the passenger compartment, the engine compartment, and the luggage compartment. Each has its own separate function, but they all have the additional function of keeping the passengers relatively safe in a crash. Two examples of frames are shown on the next page: the first, a space frame built by a college student in a vehicle-design program and the second, a space-frame outline generated by a computer. These would not fit the requirements for your challenge, but you can use them as guides for brainstorming. Discuss them with your team and write down at least two observations about each design.

## Objectives

- To define design factors that improve passenger safety

- To explore how engineers make safer vehicles

## Materials

- Computer access

- Copies of *Brainstorming* and *Mind Mapping* for each student

- Pencil and drawing paper for each student

**Student-Designed Model Formula Racing Frame, Engineering Technology Department, Western Washington University**

_____

_____

_____

**Computer-Generated Model of a Space-Frame Vehicle**

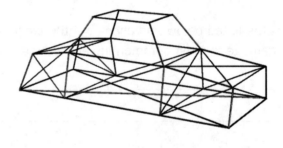

_____

_____

_____

3. What part of the frame needs to be the strongest?

_____

_____

_____

4. Continue brainstorming. Create a list of design requirements for the space frame.

_____

_____

_____

_____

_____

_____

5. The National Road and Motorist Association (NRMA) website has some suggestions on how to build strong vehicles. Visit that site at _www.nrma.com.au_ and identify two ideas that would be good to use when you design a space-frame vehicle.

_____

_____

_____

_____

_____

6. In your notebook make your first sketch of a possible shape for a space-frame vehicle. Then show it to your group. If you have time, make a second sketch that is better than the first.

# Teacher's Guide to Activity 1: Passenger Safety

*Using the Web and other available resources, students investigate the factors that lead both to vehicle accidents and accident-induced injuries. Organization websites allow students to explore the work of professional engineers. Then students should conduct brainstorming sessions to determine how to translate the acquired information into vehicle design. Students will identify factors that need to be taken into consideration in order to build a strong frame.*

## Conceptual Development

Through Web research students should observe that damage in automobile crashes occurs mostly to the front end of the vehicle. Students should also find information to support the need to keep occupants strapped into a vehicle for greater safety.

Students begin to think about what is necessary in building safe vehicles that offer maximum protection to the passengers in the event of a crash. Eggs, used as substitutes for passengers, demonstrate the rather vulnerable condition of human passengers in real vehicles.

At this point in their conceptual development, students' understanding of crash forces is intuitive. Student drawings will be rough, and students may copy a design from sources that were provided. Students will need encouragement to improve their sketches with each iteration.

## Extensions

Take a field trip to a wrecking yard or a body shop. Have students take pictures of crashed vehicles or vehicle frames. Ask each team to discuss one picture and speculate on the type of crash forces that caused the damage.

# Activity 2:
# Force-Path Diagrams I

*Question: Where does the stopping force go?*

## Procedure

In your research you saw the force of a crash on the front end of a car. To keep these forces from injuring passengers, frames must be designed to absorb the forces.

Work together so that every team member understands how to draw force-path diagrams. Remember that the arrows you draw in these diagrams represent forces acting on the structure (on the wooden pieces of your space frame). Remember also that these forces can shatter the frame and injure the passenger.

1. Imagine a 2-N force acting on the front-end portion of a space frame as shown in the diagram below. The force is represented by a 2-cm arrow; it is drawn to the scale 1 cm = 1 N. The force may have some effect on the diagonal piece too. Think of the frame from top to bottom. To indicate how much force you think is directed along each of the pieces of the space frame in this diagram, draw arrows to scale next to the individual pieces, and label each with your estimate in newtons.

**2-N Force Acting on Front End**

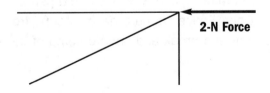

**2-N Force**

### Objectives

- To identify where stopping forces act on a vehicle

- To recognize that some forces are transmitted and others cause destruction

- To diagram the forces in a crash

### Materials

- Copy of *Work as Mechanical Energy* for each student

- Pencil and ruler

2. Now imagine Space Frame 1 crashes. In three places 2-N forces hit the front end. (The scale is 1 cm = 1 N.) Estimate how these forces will be transmitted through the vehicle. Label your estimates in newtons.

**Space Frame 1—Top View**

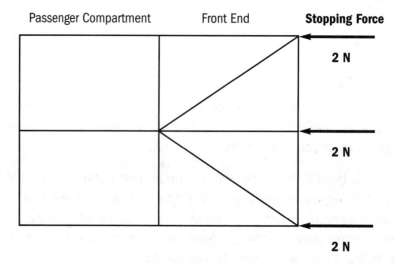

3. The forces of the crash are much greater than 2 N. Do you think the piece of wood in the center will be able to withstand that much force, or do you think it will break?

_____

_____

If the frame breaks, some of the force will be used in the crash. Estimate how much force would reach the passenger.

_____

4. Describe how well you think Space Frame 1 would perform in a front-end collision. Does Space Frame 1 have a "soft" front end; that is, would the front end break and absorb some of the stopping force?

_____

_____

Would the passengers be safe? Would enough stopping force be used up?

_____

_____

5. Look at Space Frame 2 in the diagram below. A 6-N stopping force is acting at a single point (1 N = 1 cm). Indicate how you think the stopping force will affect the structure by drawing force arrows next to each of the structural pieces and labeling them with the magnitudes of the forces involved. Show how much force you think will reach the passenger compartment.

**Space Frame 2—Top View**

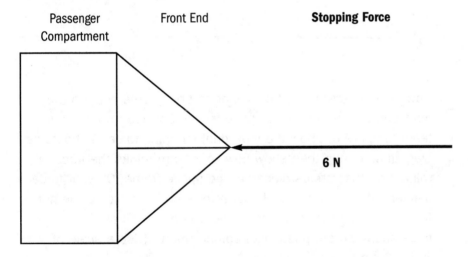

6. Describe how well you think Space Frame 2 would perform in a front-end collision. Does Space Frame 1 have a "soft" front end; that is, would the front end break and absorb some of the stopping force?

_____

_____

Would the passengers be safe? Would enough stopping force be used up?

_____

_____

# Unit 3: Activity 2

7. Choose either Space Frame 1 or 2. Can you make the front end absorb more force? Draw your ideas for the improved front-end structure here.

**Improved Space Frame—Top View**

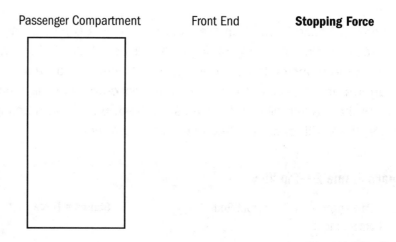

Passenger Compartment        Front End        **Stopping Force**

8. Now imagine that a total of 6 N of stopping force acts on the front end of your improved space frame. Brainstorm with your team to figure out how the individual stopping force(s) should be diagrammed, and then show how your team thinks the force(s) will act on the various pieces of the space frame. On your force diagram remember to label your arrows and to indicate the net force acting on the front end of the frame as well as the net force acting on the passenger compartment. (Use a scale of 1 N = 1 cm.)

9. In technical language describe how your team's improved space frame performs in a front-end collision.

_____

_____

_____

10. In your notebook write a description of the process you used to make force-path diagrams.

# Teacher's Guide to Activity 2:
# Force-Path Diagrams I

*Students explore the idea of force path by drawing force arrows to describe how they think stopping forces affect a space frame. Students are asked to analyze two specific designs that have been chosen to illuminate the idea of reducing the impact of forces on the passenger compartment. In this activity they consider the distribution of force, top to bottom; in the next activity they will examine the side-to-side distribution of forces. Students imagine whether the front of the frame could be crushed, and they are asked to consider how this would affect the safety of the frame. They finish by making a sketch of an improved frame design. See the following samples of possible students responses, both correct and incorrect, for force-path diagrams.*

## Sample Student Force Diagram with Labels

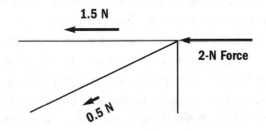

## Space Frame 1—Top View, Sample Student Answer

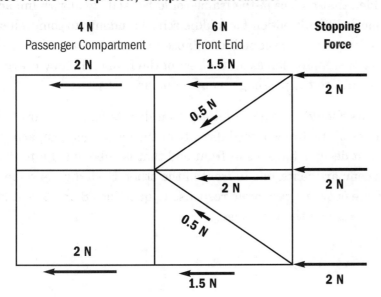

**Space Frame 2—Top View, Incorrect Student Answer**

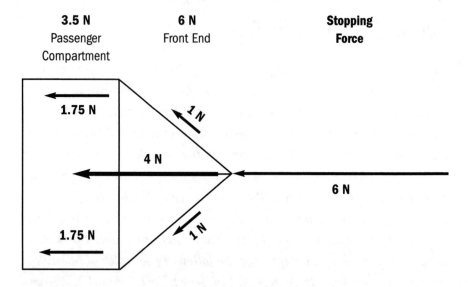

## Conceptual Development

The idea of force or load path is traditionally developed only in a college-level engineering course, but students intuitively understand the idea. Although they lack the mathematics necessary to calculate angular components of force, they can estimate these vectors with reasonable accuracy and show them graphically. In this activity they look at forces along the top and bottom of the vehicle. Make sure that they label their force arrows in newtons, not centimeters.

The idea that a piece of the frame could be "soft" to absorb impact is a more difficult notion for middle school students to comprehend. A soft front end is designed to collapse in a collision to reduce the impact of stopping forces on the rest of the frame, thereby preventing damage to the passenger compartment.

Students should sketch a front end that directs force outward, with no pieces of the frame aimed directly at the passenger compartment. The best designs have a soft front end that is subdivided into two or more compartments, one in front of the other. Each directs some of the force outward. A sample response to questions 8 and 9 is shown in the figure on the next page.

**Improved Space-Frame Student Sample—Top View**

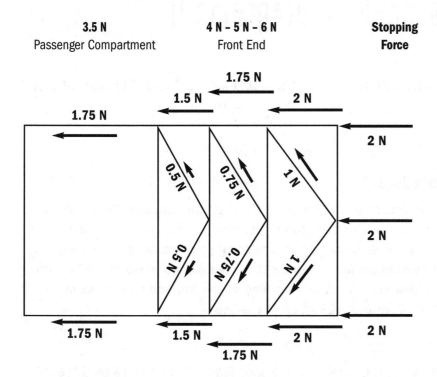

Later activities develop the concept of work as a force acting over a distance. Students learn that there is a trade-off between stopping distance and stopping force. They can understand that if a very rigid object comes to a sudden stop over a very short distance, the impact, or force that is sustained, is greater than if an impact is distributed through a soft front-end structure.

## Extensions

Most American automobiles have impact-absorbing plastic connectors attaching the front bumper to the car. Students can look for these under the front ends of their own vehicles. To be sold in the United States, automobiles must be able to withstand a low impact without forces being transmitted to the passenger compartment.

If a passenger is not firmly attached to the car chassis, the same forces that stop the car will impact the human. An air bag is a crushable structure that performs the same function for the passenger that the crushable bumper does for the vehicle. Students can research air bags on the Web.

# Activity 3:
# Force-Path Diagrams II

*Question: Where does the stopping force go along the sides of a car?*

## Objectives

- To examine stopping forces that act on a space frame

- To represent stopping forces with diagrams

- To design improved vehicles that absorb stopping forces

## Materials

- Copy of *Work as Mechanical Energy* for each student

- Pencil and ruler for each student

## Procedure

In this activity you will explore a design that causes the stopping forces acting along the top of the front end to be transmitted to the wooden wheel block. When designing your actual space frame, you will need to decide where most of your pieces of wood will be—on the sides such as in this activity, on the top and bottom as in Activity 2, or a combination of the two.

1. Imagine the side view of Space Frame 1. In a collision three 2-N forces act (1 cm = 1 N). Draw force arrows along each of the individual structural pieces to indicate the forces that you think are acting on the space frame's front end and on the passenger compartment. Label each arrow with your estimate of the size of the force in newtons.

### Space Frame 1—Side View

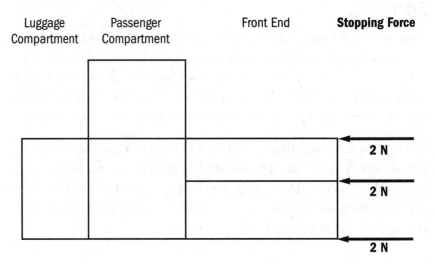

2. In technical language describe how you think Space Frame 1 would perform in this front-end collision. Describe how well you think Space Frame 2 would perform in a front-end collision.

_____

_____

_____

Would the passengers be safe? Would enough stopping force be used up?

_____

_____

3. Consider Space Frame 2 from the side. Show how you think the stopping forces would be transferred by the front end into the passenger compartment. Label the arrows with the size of the force (1 N = 1 cm). Also, show the total force that is transferred into the front end and the passenger compartment.

**Space Frame 2—Side View**

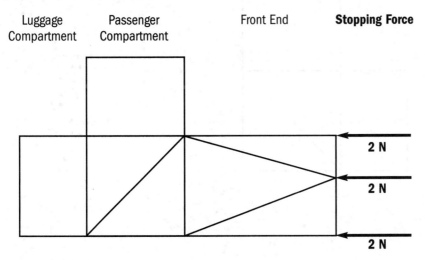

4. How would Space Frame 2 perform in this front-end collision?

_____

_____

Would the passengers be safe? Would enough stopping force be used up?

_____

_____

5. This picture shows the side view of a space frame with only a passenger compartment and a luggage compartment (no front end). As a team, design an impact-absorbing front end and then draw a force-path diagram showing a 6-N stopping force acting on it.

**Space Frame Without Front End—Side View**

| Luggage Compartment | Passenger Compartment | Front End | **Stopping Force** |

6. Explain why your team designed the front end as you did.
   Describe how you think your front end would perform.

   _____

   _____

   _____

   _____

   _____

   _____

   _____

   _____

7. In your notebook review what you have written about force-path
   diagrams, and add what you have learned in this activity.

# Teacher's Guide to Activity 3:
# Force-Path Diagrams II

*Students apply what they have learned in Activity 2 to the sides of a space frame. A luggage compartment is added to the sample designs. The hardest part of this activity may be the change of frame of reference—imagining a space frame from a side view.*

*The two sample space frames should lead students directly to an effective impact-absorbing ("soft") front-end design. With practice students should be able to design a very good front end and make a good force-path diagram for it in the last section of the activity. In Activity 4 each team will have to decide whether to build a top-and-bottom structure or a left-right structure.*

## Conceptual Development

There are at least two new ideas that may emerge from this activity. Students may recognize that diagonal structural elements tend to break more than horizontal or vertical elements. The alignment of the structural elements is one of the differences between Space Frame 1 and 2. This design consideration is explored further in Activity 4, but since students will probably be aware of the alignment issue at this point, many of their designs should have "V" structures with the open end of the "V" facing the rear of the car. Also, they may realize that the stopping force should be transferred downward, to the wheel block, as demonstrated in the diagonal brace across the "door" of Space Frame 2 (see diagram on next page). These observations can be encouraged with good dialogue as teams discuss potential designs: "Where are the stopping forces?" "How can they be redirected?"

Students should realize that the 2-N forces acting at the top and bottom of the frame will transfer the stopping force right through the front end. You may need to explain that the center stopping force will break the vertical piece of the passenger compartment, thus transferring little force. Point out to your students that the top force is transferred right through to the luggage compartment, causing the rear end to break. Sample design solutions are presented in the following diagrams. Keep in mind that any reasonable solution suggested by students should be accepted.

## Space Frame 1—Side View, Answer

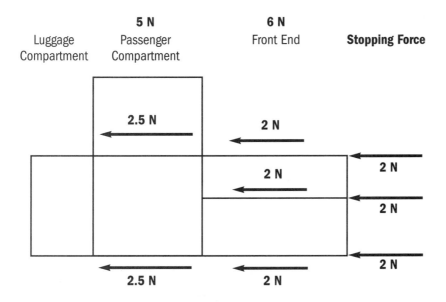

Have students consider the places where the 2-N straight force meets the 1-N diagonal force. They should predict some breakage there, with some force transferred to the passenger compartment. Then point out the diagonal across the "door," and explain to them that the diagonal piece will carry the force to the wheel block, which is not shown in the figure.

## Space Frame 2—Side View, Answer

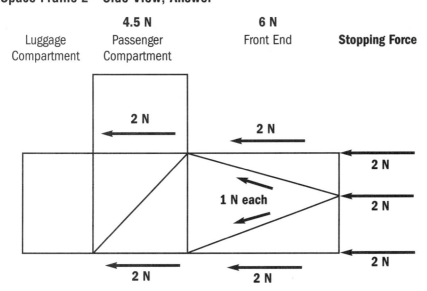

Encourage students to maximize the front end and minimize the size of the passenger and luggage compartments. Some students may realize that the luggage compartment could be placed in front of the passenger compartment. Discuss the accordion effect of some designs to absorb impact. In general, V-shaped structures break more easily than arched structures.

**Space Frame—Side View**

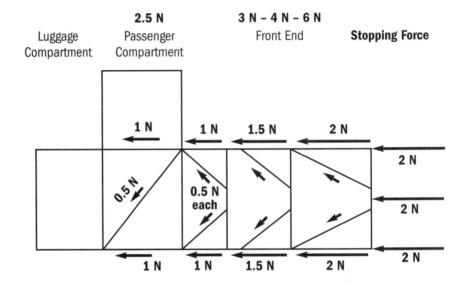

## Extensions

Ask students to find examples of accordion-like safety features in cars and on the roads. Students can look into the engine compartments of various cars and see the collapsible steering columns. They can see the yellow barrels of sand or water that sit on freeways in front of concrete hazards. Ask them to take photographs or digital images of the safety features they find and share them with the class.

# Activity 4:
# Testing Material Properties

*Question: How can the properties of materials be compared?*

## Procedure

Once car bodies were made of wood. Today most are made of metal, but plastics are being used more often. Wood is a natural material. The strength of wood depends on its type, the conditions under which it was grown, and how it was prepared. Plastics are polymers made from oil. They vary in strength and flexibility. Metals are usually various elements combined as an alloy to meet specific requirements. Metals are very strong but heavier than wood or plastic.

Comparing the properties of materials helps engineers write design specifications. In this activity you will compare materials that could be used in your space frame. In the testing of materials, test pieces must be the same size, and samples must be tested in the same manner. In this way you will get results that can be meaningfully compared.

1. Your first task is to create a chart in which you will record the results of each test (see sample chart below). Place the chart in your notebook.

### Sample Materials Testing Data Table

|  | MASS | BENDING STRENGTH | TENSILE STRENGTH | ELASTICITY |
|---|---|---|---|---|
| Dowel Rod |  |  |  |  |
| Balsa Stick |  |  |  |  |
| Plastic Rod |  |  |  |  |

## Objectives

- To test a variety of materials for strength and flexibility

- To determine which material will best meet the requirements

- To determine the best method of attaching parts

## Materials, per team

- Glue (at least 2 brands)

- One stick of balsa—3 mm square and 30 cm long

- One birch dowel rod—3 mm square and 30 cm long

- One plastic rod—3 mm square and 30 cm long

- Handsaw (with optional miter box)

- Two small clamps

- Four sticks of balsa—3 mm square and 6 cm long

- One small bucket with handle (e.g.: child's sand bucket)

*continued on next page*

**Materials** *continued*

- About 0.25 kg sand

- 2.5-N spring scale

- 1-kg mass with hook

- 5-kg mass with hook

- Triple-beam balance scale

- Eye protection

- Copy of *Student Safety Guide* for each student

2. Obtain two kinds of wood and one kind of plastic rod. Each should be exactly 30 cm long. If they are not the right length, carefully cut them.

3. An important aspect of selecting materials for vehicle construction is determining how much a specific amount of material will weigh. Some materials like concrete are very strong but also very heavy. Other materials like plastic and aluminum are light but not very strong. In vehicle design, engineers search for the lightest material that will do the best job at the most reasonable cost. Weigh each test piece with a triple-beam balance and record the weight of each test piece in the appropriate space on your chart.

4. A material's ability to resist being bent is referred to as its bending strength. A piece of material with high bending strength is also referred to as being stiff. Clamp one end of the test piece to the top of a table. Allow only 2 cm of the piece to overlap the table. Measure the distance from the other end of the test piece to the floor. Attach the hook of a 1-kg mass to the other end, 2 cm from the end. Place a small piece of tape on the top of the hook so it will not slide off. Allow the test rod to bend, and measure how far it is to the floor. Subtract from the first measurement to find how much the rod bent.

5. The ability of a material to resist being pulled apart is referred to as its tensile strength. For example, nylon string has greater tensile strength than cotton sewing thread. In this project you will need to join pieces of wood with glue. Compare the strength of glue joints in this way: Carefully cut two 6-cm sticks of balsa wood on angles, so that there will be a good surface to join. Join them together with wood glue. Do the same thing with hobby glue. Make sure the angles are the same. Allow the glue joints to dry overnight. (Go to the next test and come back to this test during the next class.) Clamp one end of each glued rod to the table, leaving a 2-cm overlap. Attach a sand bucket to the other end, 2 cm from the end. With one member of the team ready to catch the bucket, slowly add sand to it until the glue joint breaks. When it does, weight the bucket and the sand and record the result on your Materials Testing Data Table.

6. The ability of a material to bend and to return to its original shape is referred to as its elasticity. When we are using a lever arm, or a support beam, the material may bend, but it must return to its original shape. Follow the steps for bending strength (step 4), but this time use a 5-kg mass. Leave the mass on the end of the rod for five minutes. Remove the mass and then measure the distance again. Did the material return to its original position? Enter either "yes" or "no" in the appropriate box on the chart.

7. With your team, review the results of your materials tests. To make your space-frame vehicle, you want a material that is light but strong. Your material should be flexible, and the joints should have good tensile strength. Discuss with your team which material and which glue you will use to construct your vehicle. In the space below, write what you have decided and why you chose those materials.

_____

_____

_____

_____

_____

_____

_____

_____

_____

_____

_____

# Teacher's Guide to Activity 4:
# Testing Material Properties

*Students will have a limited variety of materials from which to choose for the construction of their frame prototype. They will learn appropriate procedures for testing the various characteristics of the materials under consideration, and they will use the results of such tests to determine which material will best meet the goals and functions of their design. Finally, students will select the construction methods that they feel would be best suited to their design.*

*The wooden sticks used for this exercise are commonly found in hobby shops and in the discount builders' supply depots. If time and budget permit, expand the comparison to more than two kinds of wood. If you have difficulty getting plastic of comparable sizes for each team, you can substitute swizzle sticks. Make sure to modify the length of all of the rods to the size of the plastic so that the testing will be consistent. Students should wear eye protection during testing because pieces can break and fly.*

*You can also compare different types of glue. Take some time to explain to students how to cut the sticks on a 45-degree angle so that there will be a good surface to join. In their final designs the positioning of the angle will influence the strength of the vehicle. Make sure that Material Safety Data Sheets are available for all the glues you bring into the classroom and watch for vapors that may bother sensitive students.*

## Conceptual Development

Middle school students are just beginning to understand the principles of a controlled experiment or "fair test." This series of activities considers one variable at a time, comparing three or more potential construction materials. The activities also compare two or more types of glue.

Reinforce the idea that only one variable may be changed at a time by asking students, "Is everything else the same?" as they prepare their results. If students suggest other safe variations to the tests, encourage them to explore a little further.

This exercise also supports the use of data in decision making for the design process. Make sure that students maintain an organized summary of their data, and refer to it as you help them through the manufacturing process. Moving from "I think…" to "The data suggests…" is another step in the growth of logical reasoning in early adolescents.

# Activity 5:
# Design Documentation

*Question: How can you use graphics to document your design?*

## Objectives

- To use technical graphics to develop design

- To clearly communicate design details

- To use technical sketches in construction

## Materials

- Copy of *Sketching Guide* for each student

- Grid paper

- Birch dowel rods or balsa sticks for each team

- Woodworking glue

- Hobby glue

- Energy-absorbing materials such as foam

- Egg (passenger) compartment made from egg carton for each team

- One wheel-block assembly per team

## Procedure

In this exercise your team will develop a design for your vehicle. It must meet these requirements:

- The frame must strap to a wheel block (flat wooden cart) of the following dimensions: 14 cm long, 8.8 cm wide, 2 cm thick.

- The passenger compartment must be 7 cm long, 6 cm wide, 8 cm high and must have a clear windshield opening at least 4 cm wide and 3 cm high.

- The roof needs an opening at least 3 cm wide and 4 cm long.

- The minimum dimensions of the luggage compartment are as follows: 3 cm long, 6 cm wide, 4 cm high.

- The maximum height of any point on the front end is 5 cm.

- The maximum overall size is as follows: 28 cm long, 8.8 cm wide, 9 cm high.

1. Finalize the frame design. Review the ideas and sketches you have made for the design of your frame, and then pick the best design.

2. Make a technical sketch of the final design, showing the frame dimensions as well as the position and orientation of each of the frame components. (Use grid paper and draw your diagram to scale.) Show both the side view and the top view. Use a scale of one grid line per centimeter. See the figure on the next page for the proper arrangement of these two views as well as the top view. (This is not the best design. It is only an example.)

**View Alignment**

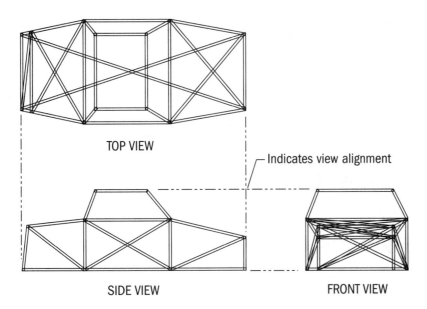

TOP VIEW

— Indicates view alignment

SIDE VIEW       FRONT VIEW

3. Using your drawings as a guide, cut out the parts of the frame using a handsaw and miter box (optional). Use tools only in the location designated by your teacher and wear safety glasses.

4. Glue the frame together carefully, one joint at a time. Think about how the forces will act in a crash as you assemble each joint.

5. Construct a seat and a restraint system for your passenger (the egg). You can use foam or cushioning material. Then strap the passenger into the vehicle. (In real cars, the restraining device is a seat belt.)

6. Fasten the wheel block to the frame. You are now ready to subject the vehicle to a crash test.

# Teacher's Guide to Activity 5:
# Design Documentation

*Students will finalize the design of their frames and make technical sketches documenting their designs. They will select from available wood for the construction of their test prototype. Students must be aware of both the need to protect the passenger and the importance of maximizing the performance and efficiency of the vehicle. The testing must take into account the mass of the vehicle as well as how well it can protect the operator and the passengers.*

## Conceptual Development

Students will learn to translate ideas and information into working drawings through technical sketches. They will learn to interpret these sketches while constructing their prototypes.

As construction continues, refer students back to their test data and their force-path diagrams. One key to successful tests is joint construction; knowing how the impact will compress the joint can help students make appropriate angles and solid connections. Encourage students to rely upon their data to inform each decision in the design process.

## Extensions

Today's computer-assisted drawing (CAD) processes are integrated with computer-assisted machining (CAM) in CAD-CAM operations. Many high schools have CAD-CAM technology stations. If possible, have one of the designs from this unit transferred to CAD and allow students to observe the machining process. Students also could attempt to create frame designs using 3-D modeling software (for example, AutoCad Lite, 3-D Studio, or Rhinoceros).

# Activity 6:
# Work and Energy

*Question: How is the amount of work needed to push a vehicle up a ramp related to the amount of work needed to stop the vehicle?*

## Procedure

In this exercise your team members will assume the roles of scientist, engineer, technician, and coordinator. The scientist needs to focus on the concepts of work, energy, and the force-distance trade-off. The engineer needs to take good data. The technician assures that each process is conducted accurately, and the coordinator keeps records and manages time.

Some of the most important concepts explored in this activity are energy and work—and the relationship between the two. Energy is the ability to do work. Work is force acting over distance. The symbols for the concepts are used in the figure below. Keep these definitions in mind and in your notebook as you follow these steps.

**Force and Distance Symbols Used in Activity 6**

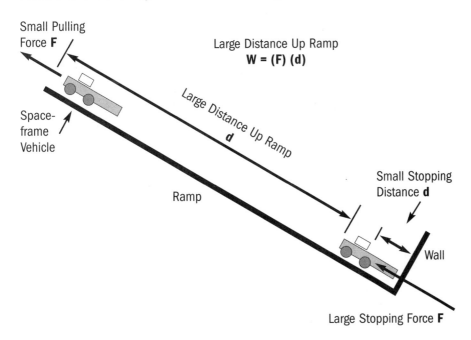

Small Pulling Force **F**

Large Distance Up Ramp
$$W = (F)(d)$$

Large Distance Up Ramp **d**

Space-frame Vehicle

Ramp

Small Stopping Distance **d**

Wall

Large Stopping Force **F**

## Objectives

- To define work as a force acting over a distance
- To describe energy as the ability to do work
- To calculate the amount of work needed to pull an object up a ramp

## Materials

- Completed space-frame vehicle
- Egg (passenger)
- 2.5-N spring scale
- Meterstick
- Ramp
- Copy of *Work as Mechanical Energy* for each student

# Unit 3: Activity 6

1. Put your passenger (egg) in your vehicle. Strap it in. Using the 2.5-N spring scale, pull your completed space frame vehicle up the ramp. Keep the spring scale parallel to the ramp, so that what you are measuring is the component of the force that acts along the direction of motion of the vehicle. Record the size of the pulling force in the space provided. (Remember inertia. The force will increase at first and then remain steady. Record the steady force.)

$$F = \underline{\hspace{2cm}}$$

(You may remember that the net force includes the pull you are providing minus the opposite force of friction on the moving parts of the vehicle. In this exercise, the wheels must move the same distance up and down the ramp so friction is the same each way. For that reason we can ignore the force of friction in our crash tests.)

2. Measure the distance that your vehicle actually moves up the ramp by starting with your vehicle at the bottom of the ramp. Measure the distance to the top of the ramp. This is the distance over which the pulling force acts. Record this distance in the space provided.

$$d = \underline{\hspace{2cm}}$$

3. Calculate the total amount of work needed to get your vehicle up the ramp. Write the measurements of force and distance in the appropriate parentheses below. Be sure to use the correct units. To calculate the amount of work done on the vehicle, multiply and record the result. Your answer should be expressed in units of joules (J). A joule is one newton-meter.

$$W = F \times d = (\underline{\hspace{1.5cm}}) \times (\underline{\hspace{1.5cm}}) = \underline{\hspace{2cm}}$$

4. When your vehicle is sitting at the top of the ramp, where is the work you did to pull it up there? Work is energy, and energy is never lost, only transferred to other forms. (HINT: Consider what will happen when you let your vehicle go.) Where is the energy of the vehicle?

_____

_____

_____

5. When your vehicle crashes, the front end may collapse. That would absorb the energy of the vehicle. Measure this distance, and record the value in the space provided. (You must change centimeters to meters so that the units will be the same as in step 3.)

$$d = \text{_____}$$

6. Recall that work was done in pulling your vehicle up the ramp. Discuss with your team whether you think that the amount of work needed to pull your vehicle up the ramp is just about the same as the amount of work needed to stop it after it rolls back down. Describe the results of your discussion on this issue, and explain the reasoning that was used in reaching a conclusion.

_____

_____

_____

_____

7. Even though there are many small variables in real-world problems, we can usually get good information about motion by making some assumptions. In this activity assume that the work to pull your vehicle up the ramp is about the same as the work to stop it. (That means that the effect of friction on the wheels and axles is very small.) Assume that all of the force is absorbed in crushing the front end of your vehicle. Then you can calculate a theoretical stopping force. Divide the work by the distance, and record the quotient in the space provided. The answer, which should be given in newtons (N), is the theoretical stopping force.

$$(F) (d) = W$$

$$F = W / d = (\quad) / (\quad) = \underline{\hspace{3cm}}$$

8. Remember the work done pulling the vehicle up a long distance will be the same as the work done stopping the vehicle over a short distance. So a big distance times a small force will equal a small distance times a much larger force. Does the value calculated for the stopping force seem reasonable?

_____

_____

9. In this calculation we have assumed that the design of the vehicle will be perfect and all of the force will be absorbed by the front end. But what if it wasn't? Where else could the force of the crash go?

_____

_____

10. Remember to enter the concepts of work and energy in your notebook.

# Teacher's Guide to Activity 6:
# Work and Energy

*Students measure the force it takes to pull their vehicles up the ramp and the distance over which the force acts to calculate work. Student teams discuss the idea that this work is stored as potential energy. This energy is transformed to kinetic energy as the vehicle rolls down the ramp. Teams use this knowledge to calculate a theoretical stopping force based on a stopping distance equal to the length of their vehicle's front end.*

## Conceptual Development

Students may never have considered the notion of work as it is defined and understood by scientists. You may need to devote class discussion time to this concept. The concept that energy (work) is conserved (work going up equals work going down) is an important one.

Some students may have difficulty with the assumptions that simplify the calculations: that friction is small and that all of the force is absorbed by the front end of the vehicle.

The calculations can show that a relatively small force over a relatively large distance (the ramp) becomes a large force over the short distance of the car's front end. Another way to present the idea of conservation of energy (work) is to review the idea of a ramp as a simple machine. Assume that a ramp is 1 m high but 2 m long. The work done lifting the vehicle straight up will be equal to the work done pulling the vehicle up the ramp, but the force needed on the ramp will be approximately half as much (if friction is negligible) $(F)(d) = (F)(d)$. Once students understand the idea of conservation, they can apply the same concept to the collision process.

## Extensions

Under normal circumstances vehicles stop because their brakes—not a wall—absorb the stopping forces. Students should explore this idea: The greater the kinetic energy of a moving vehicle, the greater the distance required for the brakes to work. More massive vehicles need correspondingly greater stopping distances.

Have students go back to the websites listed in Activity 1 and estimate the stopping distances from the crash-test videos that are provided on several of them. Supply students with a value for the stopping work by using a reasonable value for the mass of the vehicle in the crash test and and then calculating the kinetic energy (KE) right before the collision using the equation KE = m v2/2. Students can then calculate theoretical stopping forces for actual crashes.

# Activity 7:
# Destructive Testing and Analysis

*Question: How safe is a space-frame vehicle?*

## Procedure

The final stage of the design process is performance testing. In the case of our space-frame vehicles, the test will include a crash. If the impact is absorbed by the front end of the vehicle, it will be destroyed, but the passenger (egg) will be safe.

1. Secure the vehicle to the wheel block.

2. After the wheels are attached, place the egg in the passenger compartment and secure the passenger with the restraining device that you designed for this purpose. Measure and record the total mass (wheel block plus vehicle plus passenger).

_____

3. When it is your team's turn, place the vehicle at the top of the ramp and let it roll.

4. Analyze the results by answering the following questions:

   ▪ Did the egg survive or break?_____

   ▪ Did the vehicle break, and if so, where? _____

   ▪ If the vehicle broke, did this help the egg to survive? _____

   _____

## Objectives

▪ To measure the effect of a crash on a vehicle

▪ To analyze the results of a crash test and determine the cause of vehicle failure

▪ To make recommendations for redesign to increase safety

## Materials

▪ Track or ramp for vehicle crash test

▪ Completed space-frame vehicle

▪ Egg

▪ Spreadsheet for recording test results

▪ Triple-beam balance scale to measure vehicle mass

# Unit 3: Activity 7

5. Record the results of the crash and your analysis of the crash events.

_____

_____

_____

_____

_____

6. After you have analyzed the results of the crash, make a technical sketch of a redesigned vehicle frame that would increase the safety of the passenger.

# Teacher's Guide to Activity 7:
# Destructive Testing and Analysis

*Students will test the vehicle prototype by crashing it into a solid barrier. Then they will analyze the results and suggest redesigns that might improve both passenger safety and vehicle efficiency, using either different materials or different configurations of materials.*

## Conceptual Development

Students learn to conduct tests and to record the results. Middle school students will tend to abandon their vehicles once they have crashed. Encouraging them to carefully analyze the results is an important goal. Have the students make a new multiview technical sketch of a redesigned vehicle frame based on their analysis of the test results. Ask probing questions: "Where did the forces act?" "Where did the most destruction occur?" "Did the passenger remain in the vehicle?"

## Extensions

An analysis of the results of the crash can lead to a discussion of passenger restraint systems. Air bags minimize injury from front-end collisions but do not help with side-impact crashes. Seatbelts and air bags are both necessary for maximum safety. If you have allowed the use of foam (as a substitute for air bags) protection around the passenger eggs, students may have data to evaluate this protection. If not, you may allow some students to redesign a vehicle with "air bag" protection.

# Activity 8:
# Energy Transfers

*Question: When a car stops, where does all the energy go?*

## Objectives

- To describe a crash from the passenger's point of view

- To compare a hypothetical force path to the actual effects of force on a structure

## Materials

- Crashed space-frame vehicle

## Procedure

Your team has managed to design, build, analyze, and destroy-test a space-frame vehicle. This activity provides an opportunity to examine your vehicle's remains and discuss your experiences.

1. Carefully place your crashed space-frame vehicle on a table. In the spaces provided, sketch your vehicle as it appeared before the crash; you may need to refer to the sketches you made in Activity 6. Indicate which pieces broke by drawing them in some other color or highlighting them.

**Top View**

**Side View**

2. Consider what you have learned about force paths, and describe why you think the pieces of wood (structural elements) broke (failed) in your space frame. Did you plan for the pieces that broke to be "soft"? Explain your response.

_____

_____

_____

3. Did your vehicle protect your passenger egg?
   Explain why or why not.

_____

_____

_____

_____

_____

# Unit 3: Activity 8

4. Imagine you are the passenger (the egg). Write a description of the energy transfers that occurred from the time you started rolling your vehicle up the ramp until you picked up the pieces.

_____

_____

_____

_____

_____

5. How is the passenger egg's experience like that of a human passenger in an automobile crash?

_____

_____

_____

_____

_____

6. Review all the notes you have taken about the concepts and processes explored in this unit. Compare your notes to those of your team members and make any necessary additions. Construct a table of contents for your notebook so that it can be included in your design portfolio.

# Teacher's Guide to Activity 8:
# Energy Transfers

*Students sketch their space-frame vehicles and indicate in color which pieces broke in the collision. They then apply their knowledge of force paths to try to explain why particular pieces of their frame broke. Students are then asked to change their perspectives and describe the energy transfers from the point of view of their passenger egg. Finally, they consider the similarities between what happened to the eggs and what happens to humans in vehicle crashes.*

## Conceptual Development

Students compare their hypothetical force paths to the results of real forces acting on their space frame. This is a simple, tangible exercise that allows students to apply the theoretical knowledge they have acquired to an actual crash. To force a serious change of perspective, students look at energy transfers from the point of view of the passengers (eggs). Emphasize the concept of conservation of energy throughout the discussions. You can ask students to identify force paths on their diagrams as a final assessment.

Here is an example of a student's description of the crash from the egg's perspective:

> *Egg was strapped in and seated facing the crash wall when a hand came down and pulled his vehicle backwards up a huge ramp. Egg could feel that energy was being imparted to him. The work suddenly stopped, and Egg could see that he was very high up a ramp but still facing the dreaded crash wall. Egg was thinking about his VRT (the scientists and engineers that worked on the space frame he was counting on to save him) when, all of a sudden, the energy that he had gained in being pulled up the ramp started appearing in the form of speed. He was speeding up toward the crash wall! Then it happened! A huge stopping force acted on the vehicle as the space frame's front end hit the crash wall. Thankfully, Egg's VRT had designed a long, soft front end in order to deflect the huge stopping force away from him. They had done this so that the front end would be crushed like an accordion, thereby reducing the impact on him. Egg started to think about all the times his VRT*

*had goofed off, but then it was all over: He was sitting safely in his seat with his nose almost touching the dreaded crash wall. He was thankful that his VRT had understood that the work needed to stop the vehicle in a crash was the same as the work that it took to pull the vehicle up to the top of the ramp, so they had incorporated the longest possible stopping distance into their design in order to minimize the stopping force. While thinking nice thoughts about his VRT, Egg wondered where all that energy had gone. Then he remembered the sound of the crash and realized that sound is a form of energy. It also occurred to him that perhaps the wall and some of his space frame had heated up—and, come to think of it, heat is another form of energy. He then realized that all the work originally done to pull him and his vehicle up the ramp had been transferred into other forms of energy, and thus he was safe.*

## Extensions

Challenge students to actually measure the force of the collision. One way would be to employ a force sensor with a memory. Another would be to rebuild the vehicle and apply a measured force until it breaks, say by positioning the vehicle so that the front end is directed upward and then stacking weights on top.

If students do well with the egg story, encourage them to expand upon it and present it to their language arts teacher.

# Reading 1:
# Brainstorming

Brainstorming is a technique to help a group work together to generate ideas or solve a problem. Here are some ideas for ways your group can work together:

1. Use a large piece of paper on a wall or a chalkboard to record ideas.

2. Select someone in the group to be the recorder.

3. Have members of the group call out ideas. Don't discourage any.

4. The recorder should write every idea down. Use short words or phrases.

5. Have each group member take a turn suggesting an idea.

6. Do not evaluate or criticize ideas during the process. At the beginning, every idea is a good one.

7. Continue the process until the members of the group have contributed all their ideas, or until time is up.

8. As a group, consider each of the ideas in turn. Underline those that need more study. Cross out those that are impossible.

# Reading 1

Here's another technique for brainstorming in a group. This process begins silently. Follow these steps:

1. Give each person in the group a pencil and a pad of post-it notes.

2. Have each member of the group record ideas on the post-it notes.

3. When time is up, have two people put their ideas on a board or paper.

4. Group the ideas together if they are related.

5. Now have more people go up to the board and add their ideas. These group members should put their notes with notes that are similar.

6. Continue this until all groups have a chance to put their ideas up. Some groups will have many post-it notes; others will have only one or two.

7. Appoint someone to read the largest group of notes. Think of one or two words that describes how they are related. Write this word near the group of notes.

8. Repeat this process for each group of notes until you have a word or phrase to describe each.

9. Discuss each group of notes and ideas again. Decide which groups of ideas need additional study and which are impossible to do.

Whichever method you follow, remember that everyone has good ideas. Don't eliminate any suggestions until your group has had a chance to discuss them together.

# Reading 2:
# The Design Process

Technology is the development of ideas and their applications to create or improve new products and services. Through technology we solve problems, like improving automobile safety or producing new vaccines. New products help satisfy our needs for more food, better clothing, and safer homes. Technology provides things we want, like DVD players and cell phones. It also provides tools to extend our capacity—computers, for example, which help us organize, calculate, communicate, and learn better or faster than before.

Technology doesn't just happen. Some person or group designs each new product or tool. Today's engineers, industrial designers, architects, and technicians follow very specific steps in their work. This organized approach is what we call the design process.

The first step in this process is to define the problem. Designers state very specifically just what problem must be solved or what want or need it will satisfy. This is done in the form of a design brief. It includes the requirements, which might be set by a contractor or an agency like the government. More often, the requirements are defined by the consumer—what people need or want and what they will buy.

Next, designers research what others have done or thought about the problem. Technology is based upon science, so designers must understand the principles that will make the product work. The new product may combine ideas that have been used before in different ways. Today's computers make the process of searching for similar ideas much easier and faster.

Once designers have an idea, how do they know if it will really work? Usually they build a prototype, a model of the product that can be tested. The prototype can be a physical model, but it could also be a computer model or a video simulation that can be previewed by consumers. Once the prototype is built, it is tested to see if it meets the requirements.

Usually there is not one prototype but many. The design is improved again and again. Since designing is a learning process, the last step is to figure out how to improve on the design. Engineers and designers are always asking this question.

There are many solutions to each problem, but the methods are very similar. Engineers have a standard way of doing things, known as the design process. When engineers, architects, or designers solve problems, they follow these steps:

- define the problem in a design brief
- gather information
- explore ideas and possible solutions
- narrow the ideas into a single solution
- build prototypes and test them
- redesign to improve the design

As student designers, you will carry out the same steps. It is very important to remember that this process is not linear; that is, the steps will not necessarily be done in strict order. Designers may return to any step and change any aspect of the design along the way. Let's look at how this process would work for a real problem.

## The Design Brief

A design brief is a clear and concise statement of what you are setting out to do. Suppose your team was given the problem: Design a table. There are many different kinds of tables: kitchen tables, dining tables, coffee tables, end tables, drafting tables, etc. You would be confused!

Let's try again: Design a dining room table. This is better, but it still leaves too much to chance. If we design a table for twelve people, and the room accommodates a table for only six, it would be useless. The design brief must contain product requirements, which clearly define the challenge. A more complete statement might look something like this: Design a dining room table for a formal setting. The table must be modern in style, light in color, constructed of solid wood; it must seat four people comfortably and expand to seat eight. This leaves the designer room to be creative, yet it limits the goals, so that the end result will match the intent.

Requirements also might include more practical goals, such as a table that can be made in a factory, assembled by one person, and sold for under $2000. These requirements create very specific limits on the design process but help ensure that the product will be valuable to the manufacturer or the consumer.

## Research

The next step in the design process is to study what others already know about the problem. In the example above, you might look at as many tables as you can find in books and on the Web. You might interview people to discover what they like about their own tables. You might also study the materials you might use, comparing, for example, the strength, cost, and availability of woods. Finally, you might explore how tables are made to find out if new tools or methods would be needed for your table.

Very few engineering projects use only new information. Invention involves using things in different ways or applying old knowledge to new problems. For example, when engineers at Boeing design a new plane, they don't reinvent flight. They use volumes of test data about performance and materials that other engineers have already gathered. This is true with many of the products we use daily. Designers may apply new knowledge, new tools, or new materials to change a familiar product into one that is new and better.

## Idea Generation

Next, designers explore possible solutions. There are a variety of tools available to help in this process, but sketching is probably the most important. This can be done by hand or with computer-assisted design (CAD) software. Two other important tools for idea generation are brainstorming and mind mapping. You can read about these skills in other sections of this book.

At this stage it is important not to limit your ideas. Use whatever tools and processes result in many ideas, including the assistance of other people. This is one reason engineers work in design teams. Coming up with good ideas is not easy. Keep trying! You will get better as you progress. Don't stop, even when the building process begins. There is always a better way to do something!

## Narrowing the Focus

Generating ideas and gathering information are only the beginning steps in the design process. Next, the best ideas must be selected, keeping the design brief in mind. A very useful tool to aid in design selection is a requirements chart. Here's what one might look like for our table:

| REQUIREMENTS | IDEA 1 | IDEA 2 | IDEA 3 | IDEA 4 |
|---|---|---|---|---|
| Seats 4 | | | | |
| Style | | | | |
| Wood Construction | | | | |
| Expandable | | | | |
| Easy to manufacture | | | | |
| Not too expensive | | | | |
| Total | | | | |

Suppose your team came up with 50 ideas. When the truly wild ideas were eliminated, four remained. Each was then listed on a chart. The team would then discuss each requirement and rate the ideas on a scale (perhaps 1 to 10). This step might require some additional research. Then the idea with the highest total would be selected for further development. You might even add a new column to the chart if the best idea turned out to be a combination of several other ideas.

The next stage of design development is to work out the details of the design, or specifications. Design development will include decisions such as determining the following:

- how big to make things
- what the different parts of the design are
- what each of the parts looks like
- what each of the parts is made of
- how each of the parts is fabricated
- how the parts fit together

All of these issues and more will be decided during the design development stage. There are over four million parts in a Boeing 777, and each of those decisions had to be made for each part. That may be the reason why it took four years to design that airplane. If you ever fly in one, you will hope that the engineers designed each part very carefully!

## Prototype Construction

One result of design development will be technical drawings. These contain all the information you need for the prototype. In some engineering processes the first prototype is a virtual device; Boeing designed, built, and conducted virtual tests for most aspects of the Boeing 777 by computer before a physical model was ever made. Students usually build a real prototype. If the design development has been done well, all you will need to do is follow the instructions found in the technical drawings.

## Testing and Redesign

The prototype must be tested to see if it works the way it was intended. The results are measured against the requirements. In the example of the dining table, we would ask:

- Were we able to make it from solid wood?
- Does it comfortably seat four people?
- Can it be expanded to seat eight people?
- Can it be made efficiently?

At this stage, new questions might be raised: Does the table look good? Can it be improved?

As each of the questions in the testing phase is answered, we ask a new question: "How can this be done better?" Making improvements to the design might require collecting additional information, doing more testing, using different materials, or being more careful in the construction. An important part of redesign is careful record-keeping.

After the tests have been completed and the design questions answered, the process usually begins again. The design brief may be modified, new ideas may be generated, selected, and developed, and new technical drawings may be made. You might make the design easier to build or more attractive. This happens in industry quite often. Once the redesign is complete, a new prototype will be built and tested to determine how much better the results will be. This process can continue indefinitely.

If the design works as intended, companies want to sell it to make money from it. So when a design is good enough, it will be documented and then manufactured. But redesign might still continue behind the scenes. Each time there is a new version of a computer program or a new model car some information from testing and from the consumer has been used for redesign.

If the design process ended when the product was sold, we would still be riding in horse-drawn wagons. Almost all of the products we use today represent vast improvements over the first design. Engineers and designers are always searching for better ways. Some day you'll probably buy a lighter, more attractive dining room table. Boeing has the next airplane in development, even though millions of people ride their current planes safely each day. Toyota Motor Company has a goal of improving their cars in some way by at least 15 percent every year. The design process never ends.

# Reading 3:
# Mind Mapping

Mind mapping is a process for generating ideas. In mind mapping you brainstorm and then communicate your ideas with diagrams. The diagrams show connections and possibilities.

To create a mind map, follow these steps:

1. Take out a blank, unlined sheet of paper and a pencil.

2. Identify a problem. This could be something you need to fix or something you want to design. Then think of a word that represents the problem, and write this word in the center of the page. (For example, in this unit you might start with the word "vehicle.")

3. Draw a circle around the word. Then think of a related word. (For example, for a vehicle the first word you might think of is "wheel.")

4. Write the related word next to the original word, draw a circle around it, and connect the two circles with a line.

5. Think of words that relate to wheel (such as "round," "circle," "tire," and "axle"), draw a circle around each of these words, and connect these circles to the circle around "wheel."

6. Continue in this fashion, thinking of words related to the ones just added to the diagram, drawing a circle around each new word, and connecting the circles as directed above. (For example, the words "turn," "chain," "gears," and "pulley" could be associated with the word "axle.")

7. Next, go back to the first level, think of words other than "wheel" that are associated with vehicle ("passenger," "chassis," and "engines" come to mind), write these words down, circle them, and connect these circles with lines to the circle around "vehicle."

8. Your words should not be random. They should show connections. When you look for solutions to your problem you need to consider all its parts. The mind map shows you many parts and how they are connected.

Suppose your team generated the mind map below to begin brainstorming about how to reduce friction in a vehicle. Each of the words on the map could represent an item that you may need to think about. Using the diagram, you can eliminate those ideas that wouldn't be related to friction (like accelerator and linkage). Then you might take those ideas that are related to friction (like wheels, brakes, or axles) and make new mind maps for those parts.

There are no strict rules for a mind map. Brainstorm and be creative. You'll be a better designer.

**Example of Mind Map**

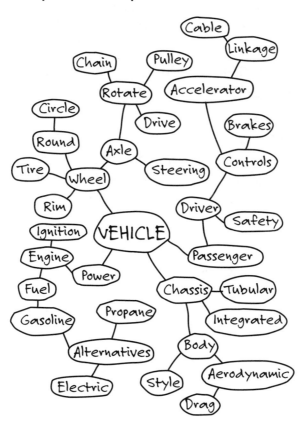

# Reading 4:
# Motion and Force

Everything in the universe is in constant motion. Just standing on Earth you are moving at more than 100,000 kilometers an hour around the Sun. The Sun is moving through the galaxy, which is moving outward in the universe. To describe an object's position we must begin from some frame of reference—the Earth, the Sun, or a point we call "start." Then we can describe the motion of an object by its direction and its speed from there.

An object at rest remains at rest or an object in motion keeps moving in the same direction until another force acts on it. This is inertia, a basic property of all matter. To cause something to move, stop, or change direction requires a force.

In the language of science and technology, distance is the space between two events, measured in meters (m), and time is the interval between two events, measured in seconds (s). Motion can be described with distance-time data tables and graphs of these data. How the distance changes with respect to time is described as speed, measured in meters per second (m/s). If we also tell the direction of the motion, we have described the object's velocity.

It is rare that only one force acts on an object. Often two or more forces act at the same time. Friction is a force that resists motion of one substance on or through another. Friction slows the motion of a whale through the ocean, or a skater on dry ice. When we combine the effects of all of the forces that act on an object, we determine the net force.

Imagine a person is walking along at constant speed, as described in the following data table and graph:

## Data Table 1 and Graph of Constant Motion

| DISTANCE (m) | TIME (s) |
| --- | --- |
| 0 | 0 |
| 1 | 1 |
| 2 | 2 |
| 3 | 3 |
| 4 | 4 |
| 5 | 5 |
| 6 | 6 |
| 7 | 7 |
| 8 | 8 |
| 9 | 9 |
| 10 | 10 |

To describe this motion, a scientist or engineer might say: "The person moved at a constant speed of 1 m/s away from the starting point for 10 s." The slant or slope of the line represents the speed. Because time increases when distance increases, we say that these two measurements are directly proportional to one another.

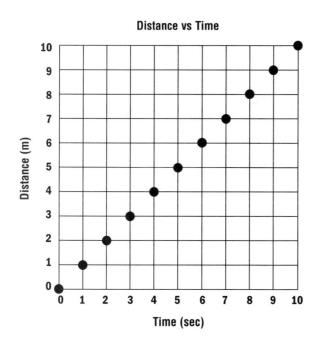

### Data Table 2 and Graph of Speeding Up

| DISTANCE (m) | TIME (s) |
|:---:|:---:|
| 0 | 0 |
| 0.5 | 1 |
| 2 | 2 |
| 4.5 | 3 |
| 8 | 4 |
| 12.5 | 5 |
| 18 | 6 |
| 24.5 | 7 |
| 32 | 8 |
| 40.5 | 9 |
| 50 | 10 |

Now imagine a car speeding up, so that every second it moves faster and faster, covering more and more distance in every 1-s interval. To do this the car would have to use fuel to generate more force than friction could subtract. The distance-time graph might look like this:

Notice that points plotted on the graph below curve upward. The slope of the graph is positive, but it is not a line. The car is accelerating.

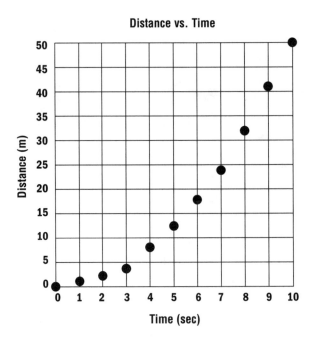

The mass of an object is the "amount of matter in it." It is an object's mass that controls its motion. The more mass an object has, the more force is required to move it. If a 1-N force acted on an object with a mass of 2 kg, the object would accelerate only half as much as a 1-kg object.

## Data Table 3 and Graph of Speeding Up a 2-kg Car with a 1-N Force

| DISTANCE (m) | TIME (s) |
| --- | --- |
| 0 | 0 |
| 0.25 | 1 |
| 1 | 2 |
| 2.25 | 3 |
| 4 | 4 |
| 6.25 | 5 |
| 9 | 6 |
| 12.25 | 7 |
| 16 | 8 |
| 20.25 | 9 |
| 25 | 10 |

In physics we measure distance, time, force, and mass to study motion. To understand how these are related, we must carefully vary only one at a time and keep all other things the same. For example, to see the relationship between mass and motion, you could change the mass but keep the force constant. You could apply a 1-N force on an object for a distance of 1 m. Then you could add mass to the object and apply the same force. Your data might look like this.

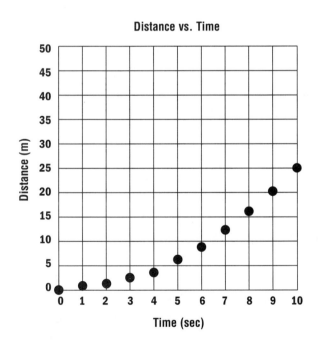

# Reading 4

## Data Table 4 and Graph of Time for Car to Speed Up as Mass Increases

| MASS (g) | TIME (s) |
|---|---|
| 100 | 0.45 |
| 350 | 0.84 |
| 600 | 1.10 |
| 850 | 1.30 |
| 1100 | 1.48 |

This relationship is important to engineers who design vehicles. Lighter cars go farther on the same fuel than heavy trucks. When students design $CO_2$ vehicles, they can use the same principles. A lighter vehicle will go farther when the force of the $CO_2$ fuel is constant.

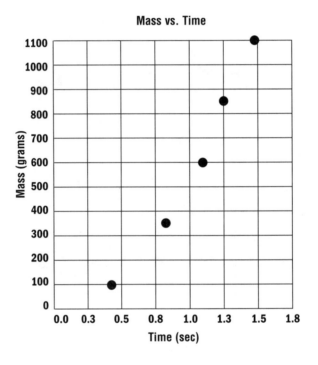

An arrow can be used as a graphic model of a force. It shows magnitude (strength) and direction. We can use the center of gravity of an object (X) to indicate the point on which all forces appear to act. It is usually the point at which the object can be balanced. In the diagram below, arrows show the frictional force $F_f$ and the applied force $F_{applied}$. Since these forces act in opposite directions, the net force $F_{net}$ is less than the force we applied. The following example shows a toy vehicle that weighs 5.9 N with a 10-N applied force and a 1.5-N frictional force. The scale of the force arrows is 1 cm = 2 N (that is, every centimeter of length of an arrow corresponds to a force of 2 N).

## Net Force Diagram

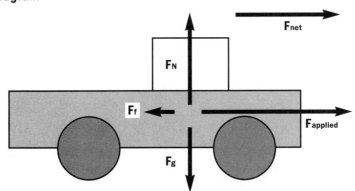

Gravity is a basic force of nature that acts on all objects. Mass is the amount of matter in an object; the force of gravity acting on an object is its weight. Earth pulls on an object, and the object pulls on Earth (much less, of course, because its mass is so much smaller). Mass is measured in kilograms (kg) and weight in the metric unit newton (N). On Earth, a newton is about 9.8 x the mass. On the Moon or another planet, mass stays the same, but weight changes.

Forces act not only on the outsides of objects but on the inner parts too. An example is the force stored inside an open mousetrap. Engineers analyze how forces act by drawing force-path diagrams. The diagrams are somewhat like the vector arrows we use for forces. But a force-path arrow is used to study stress within structures. The force-path diagram on the next page shows a space-frame vehicle carrying an egg as a "passenger." The arrows show how the forces move through a frame when it crashes. A large stopping force crushes the front end. A smaller force is absorbed by the protective material inside the passenger compartment. The egg is safe.

### Force-Path Diagram of How a Large Stopping Force Moves Through a Collapsible Front End

Only 1.5 N, the average stopping force, is transferred to the passenger section

3 N of initial stopping force acts on the front of the space-frame vehicle

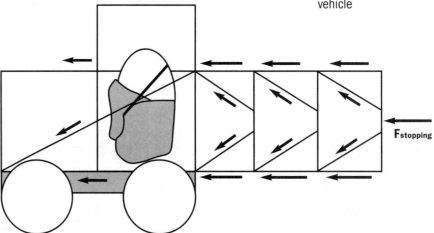

$F_{stopping}$

The study of forces provides us with information to understand and make predictions about objects on Earth and in space. Using this information, engineers can build faster and safer vehicles for today and tomorrow.

# Reading 5:
# Student Safety Guide

Have you ever seen a movie where an absent-minded scientist works carelessly in a messy lab? If you did, you know it was a fantasy. For a scientist, an engineer, a technician, or a student, safety is an important goal.

Wherever scientists and engineers work, there are special procedures that they follow. They keep their workplaces organized and secure, so that every employee is safe. In *Fender Bender Physics* you will assume the role of scientist, engineer, and technician. Your actions are important. To ensure a safe and scientific place to work, follow these guidelines.

- Be organized: Know what you are going to do, and what safety precautions you must take, in advance.

- Be informed: Know what hazards could occur. Be prepared to prevent them.

- Be cautious: Work slowly and think about what you are doing.

- Be serious: Don't let other things distract you from what you are doing.

- Protect yourself: Wear the appropriate safety equipment all the time it is needed. Clean it and put it away when you have finished your work.

- Act responsibly: There are other times for rough play or moving fast. In the classroom, act like a professional.

## Organization for Safety

For each activity in *Fender Bender Physics* you will use equipment and tools. Know what you need and how to use each tool or instrument before you begin. Read all the directions, and watch carefully when the teacher demonstrates the correct way to use a tool. Record the safety rules for each activity in your notebook before you begin, and discuss the rules with other members of your team. If someone is absent, review the rules with them before you begin.

## Get the Information

Your teacher will explain the proper ways to use tools in the classroom. This is not the time to experiment. Use a tool in the right place, and put it away as soon as you are done. Here are a few tools and materials that you will use:

- Cutting tools: Use only cutters with safety shields. Sit down and cut away from your body. Make sure everyone else is sitting and that no one is too close.

- Saws and hammers: Your teacher will demonstrate the appropriate way to use these tools and where you are allowed to use them. Saw slowly and regularly. Make sure that the wood you are sawing is firmly attached by a clamp or vise. Make sure no other students are near. Hammers can be especially dangerous if you miss the nail or tack. If you have never used a tool before, review the right procedure and start slowly and carefully.

- Glues and paints: Most of these products are toxic. Often the fumes from glues and paints are irritants for people. Make sure that the information about the product is in the classroom. Work in a well-ventilated place.

## Caution Is the Key

Your team will have one chance to make a good vehicle. If you start sawing, cutting, or gluing before you have a good plan, you will be disappointed. Review your manufacturing plan before you begin. If you must wait for the right tools, be patient. When you are cutting or sawing, take a little off at a time. You can always go back. Like all good craftspeople, "Measure twice and cut once."

## Have an Attitude

Design is serious business. It requires serious team members. Keep your conversation—and your mind—on your project. If you are distracted, you may make a mistake that could hurt your project or a member of your team.

Accidents may seem unpredictable, but they can be avoided if you plan ahead. Are people too close to the hammer? Is something on the edge of the desk? Is the glue or paint where it can be spilled? Think about what might happen, and then make sure it doesn't.

## Protective Gear

In the classroom, just like in a real workplace, protective clothing is required. Dressing for success in a lab means wearing appropriate clothing. Do not wear floppy pants or blouses or excess jewelry that might get in the way and cause an accident.

For most hands-on science projects the most important piece of equipment is safety goggles. This doesn't mean visors—and plastic lenses won't do. Safety goggles have shields all the way around and are made of unbreakable plastic. They are made for engineering work.

Safety glasses must be worn anytime you saw, hammer, or cut. They also are required anytime you use paint or other material that could splash in your eyes. Safety glasses must be worn whenever you are testing something that can break or shatter and when you crash your space-frame vehicle. When you take off your safety goggles, clean them before another student wears them. You may have a sterilizer in the room, or your teacher may arrange for you to wash the goggles in soapy water and leave them to dry at the end of class.

## Behave

In *Fender Bender Physics* you are learning not only about science and engineering but also about design. You are the designer of a new vehicle and a member of a Vehicle Research Team. Assume responsibility for your actions. In everything you do, model your behavior on that of a successful designer. Then your invention will be an outstanding success.

# Reading 6:
# Work as Mechanical Energy

Galileo first realized that it took force to change motion in the 17th century. He reasoned that if a body was moving on a completely smooth surface without any friction, it would continue forever. Isaac Newton formalized this idea into the first law of motion: If a body is at rest, it stays at rest. If it is in motion, it remains in motion in the same direction until a force acts on it. This law describes a property common to all matter, called inertia.

When a force acts over a distance, the result is work. Work is the fundamental measurement of energy. *Fender Bender Physics* explores mechanical energy and the work done when vehicles start, move, and stop. Because energy can never be lost, we can measure it by calculating how much work a vehicle does.

You may think that reading this article is work. We may refer to school work, homework, going to work, or working out a problem. But in physics, work is done only when a force acts over distance. The formula for this is $W = (F)(d)$.

Imagine pulling a toy truck with a force of 1 N over a distance of 1 m, as shown in the figure below. The work done is (1 N) (1 m). We sometimes say 1 newton-meter, but this unit is more accurately described as 1 joule (J).

**Work on Toy Truck**

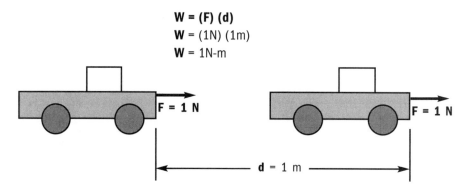

$$W = (F)(d)$$
$$W = (1N)(1m)$$
$$W = 1N\text{-}m$$

F = 1 N          F = 1 N

d = 1 m

It is easy to understand that if you pulled the truck 2 meters, 2 joules of work would be done. If the truck was stuck with glue to the ground, you could pull and pull until you were very tired, but you would have done no work at all in the scientific sense.

In order to do work, the motion and the force must be in the same direction. Here's another example: Suppose you pulled the truck on a string that was angled at 45 degrees. You would still be doing work, but only the part of the force that acted horizontally would count for the work calculation. (Of course, if you pulled too hard and the truck flipped up, work would be done in the upward direction too.) Here's another example: A waiter carries a heavy tray down an aisle. He is moving horizontally, but the tray stays at exactly the same height. The work done is equal to the force moving the waiter and the tray down the aisle, but the work does not include the effort it takes to keep the tray high in the air.

Work that is done on a vehicle changes its motion; it may speed up, slow down, turn, or stop. The work done on the toy truck in the figure on the previous page causes it to speed up (or to start moving, if it was sitting still just before the force was applied). If the force had been applied in the opposite (backward) direction, the work done on the truck would have caused it to slow down (or to start moving backward, if it was not already in motion).

To lift a toy truck, you simply apply an upward force equal to the weight of the truck. To find out how much work is done, multiply the weight of the truck (gravity's force) by the distance (height) it is lifted. The toy truck on the next page weighs 2 N and is lifted by only 0.5 m, so the work done is still 1 J (1 newton-meter).

## Work Lifting a Toy Truck

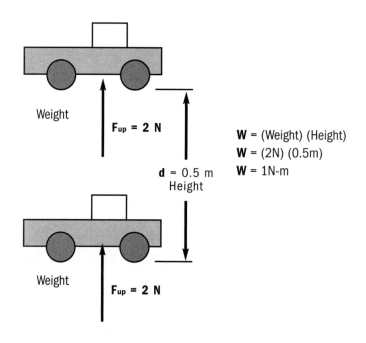

Work represents energy—something we cannot create or destroy. There are often many ways to do the same amount of work. Some work requires less force, so it is easier to do. It would take 2 N of force to lift our truck 2 m above the ground. But suppose we used a 4-m ramp to get the truck to the same height. The work (force x distance) must be the same—so if there were no friction slowing the truck down, we would only need 1 N of force. Use of the ramp doesn't reduce the amount of work, but the ramp decreases the amount of force needed to do the work. That's why ramps are used for heavy objects.

## Work to Pull Toy Truck Up Ramp

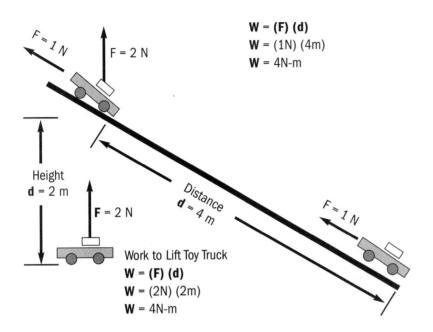

F = 1 N

F = 2 N

**W = (F) (d)**
**W** = (1N) (4m)
**W** = 4N-m

Height
**d** = 2 m

Distance
**d** = 4 m

**F** = 2 N

F = 1 N

Work to Lift Toy Truck
**W = (F) (d)**
**W** = (2N) (2m)
**W** = 4N-m

All simple machines trade distance for force to help us accomplish real jobs; pulleys, inclined planes, wheel and axles, levers, wedges, and screws move smaller forces over greater distances to do the same amount of work. In a world without friction, we can calculate the ideal mechanical advantage of a machine. Machines reduce the force needed for a job by increasing distance. Of course in the real world a very different force, friction, acts in the opposite direction to the force of the machine. So the advantage we get by using the machine is reduced.

## Force-Distance Tradeoff

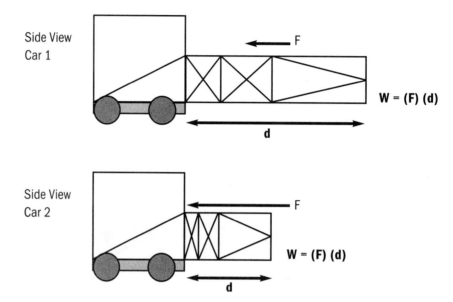

Side View
Car 1

F

$W = (F)(d)$

d

Side View
Car 2

F

$W = (F)(d)$

d

We can't create or destroy work, but we can store it or waste it. We store work when we lift an object to the top of a ramp. The object then has potential energy, which can make it come back down again. The Sun stores potential energy in water when it lifts it through evaporation. The energy is used when the water returns to Earth as rain and moves downhill in a mountain stream. Energy can be wasted too. If moving parts rub against one another, part of a machine's energy can change to heat. That energy won't be available to do work in the same machine again.

You can use your knowledge of energy and work to analyze the motion of the vehicles in *Fender Bender Physics*. If you have well-functioning wheels and axles, the friction will be minimal and you will be able to make good estimates of the work your machines will do. You can also use this to help estimate how much force must be absorbed by a space-frame vehicle in a crash. Remember to use the principles of Galileo and Newton in your team "work."

# Reading 7:
# The Manufacturing Process

The word "manufacturing" comes from two old words that mean "made by hand." Today very few of the products we use are made that way. Engineers design factories where workers and machines work together to make many products quickly.

Science and technology have worked together to make modern manufacturing more efficient and more precise. One of the first modern manufacturing procedures was developed in England in the 18th century to speed up the weaving of complex patterns. An engineer named Jacquard invented a punch card system (much like a voting ballot) that helped a machine make designs. Jacquard knits are complex and beautiful patterns that you can still buy today. Inventor Henry Ford developed the first assembly line for manufacturing automobiles in the early part of the 20th century.

Engineers design not only the products that we manufacture but also the tools that make them. Engineers determine how an object will be made and the materials that will be used. The tools are often specially made by tool-and-die technicians. Engineers plan where the workers and tools will be located in the manufacturing plant to create the most efficient movement of people and products. They also may develop software to control the mechanisms used in a factory.

Whether a product is made by hand or in a factory, the steps are the same. Once a design has been developed and tested, the manufacturer must follow these steps:

- Write design specifications: These are very specific details about how the product will be made. They include scale drawings of every part, usually done on a computer-assisted design (CAD) program, and specific descriptions of the materials. The specifications also include the order in which the product will be assembled, and the tests that will be done on it to ensure that each step has been followed.

- Prepare tools: In a hand-manufacturing procedure, craftspeople use simple tools. But most manufacturing plants use special, computer-assisted machines (CAM), which make each part exactly the way the specifications demand.

- Establish procedures: Each part of the manufacturing operation must occur in a specific order by a trained person (usually helped by a machine.) Before even one product is made, the engineers will work with the manufacturer to insure that every person involved in the process is prepared. The design engineers will also establish the speed at which the product is made.

- Check safety: In a factory safety is very important. Workers are trained to do things in a careful manner and to use equipment wisely. Each worker has specific clothing and safety equipment worn at all times. The managers monitor the lights, the power, and the air of the factory for health. The Occupational Safety and Health Administration (OSHA) advises industry on its safety standards.

## Quality Control

When a single product is made by hand, it is usually unique. But factory-made products are all alike—or are they? Look at a box of M&Ms. Is every one exactly the same size? If you had a fine balance and calipers, you would find out that they were almost the same size. There is a tiny amount of variation in everything we make.

Design engineers establish the amount of variation (tolerance) that they will allow in a product. If a candy is supposed to weigh one gram, they might allow 1/100 of a gram difference in each one. The engineers also establish points at which the variation in products will be checked. If the quality is not consistent and within established tolerance, the manufacturing process will be improved.

The purpose of quality control is to catch problems in products before they are sold. We have all heard about expensive product recalls. But they represent only a tiny fraction of the problems that manufacturers find and fix before the products are ever sold. Making a product that meets all specifications is the process of quality control. Whether a design is built by one person or by an assembly line, accuracy and safety result in a quality product.

# Sketching Guide

What does the work of Renaissance artist Leonardo da Vinci and a 21st-century rocket scientist have in common? In a word—sketching. For centuries, artists, designers, and engineers have followed a process that begins with a sketch. The seeds of a great idea—simple or complex—usually develop from a picture in an inventor's mind. But for an idea to work, sooner or later, the idea must be sketched out on the page.

Whether you consider yourself an artist, a doodler—or neither—you can improve your sketching skills. You may be confident of your skills in math and science but hesitant to share your sketches with others. This guide should help you put rough ideas on paper, so you'll be able to share them with your team. Your ideas will get better, and the construction stage of the projects you undertake will go much more smoothly.

## Begin with the Basics

Practice drawing basic shapes whenever you can. Doodling is a way of thinking out loud. Your ability to communicate with a pencil and paper can improve with repetition. So use your napkins, your scrap paper, and the margins of old papers. Whenever you can, use a soft pencil rather than a pen. Ready, set, draw.

Try circles, squares, lines. Think about basic shapes. Are the lines of your drawing parallel? Have you considered size? When you first begin to draw, your shapes will be rough and irregular. But practice makes perfect. Each basic shape will be more regular.

# Sketching Guide

When you are confident that you can draw a circle and a square, add a second dimension. Make your circle a sphere and your square a cube. Check for perspective again. Are the lines of your cube parallel? Try to add a little shading to your sphere. To do that, imagine a light bulb or the sun shining down on the sphere so you can visualize where the light is coming from and which side of the sketch should have shading.

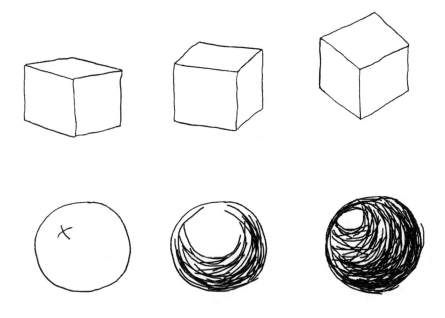

Then try an exercise combining these basic shapes. Make a stick man or a car, composed of circles, squares, and lines. It's all right if it looks like a cartoon. You are expressing your ideas in a way that others can share.

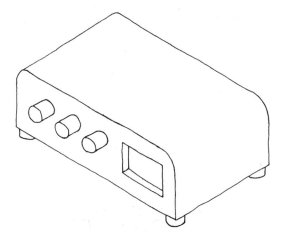

## Consider Scale

Artists can take the liberty of making things any size they want. But size is essential to a scientist or engineer. Learning to scale a drawing is extremely important.

Take a basic figure, such as a mousetrap car. Imagine that you are drawing your car on a small piece of graph paper to scale. Can you calculate how big each component will be?

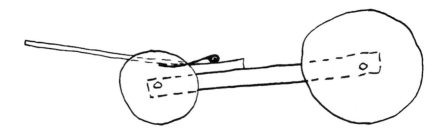

| COMPONENT | SIZE OF ORIGINAL | SIZE OF SCALE DRAWING |
|---|---|---|
| Length of chassis | 60 cm | 20 cm |
| Width of chassis | 30 cm | |
| Length of arm | 75 cm | |
| Diameter of wheels | 6 cm | |

**Answers: 10, 25, 2.**

How do you combine all this math in a sketch? Many scientists and engineers use grid paper for their sketches. This gives them the freedom to doodle—along with the accuracy of a mechanical drawing.

Can you interpret this drawing done to scale? If one box on the grid paper represents one centimeter of the true dimensions, how long will this car be? How long should the material for the chassis be?

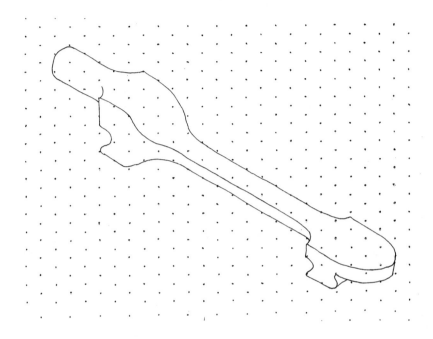

How thick should the chassis be? As you can see, drawings done to scale provide information from which engineers can develop design specifications.

## Work Out the Details

On a design team, attention to detail is crucial. An inventor may have suggested a grand idea, but carrying out that idea depends upon countless details. Each wheel, each axle, each surface, and each connection must be drawn to accurate specifications.

This process requires good communication and coordination among members of a team. Every person on a team must know a project's goals and requirements—and have a clear vision of the end product. And the dimensions must be *perfect*. Imagine a car built by a poorly coordinated team: When the wheels, the chassis, and the steering wheel come together, nothing fits! Team members who pay attention to the details prevent that from happening.

Each design team member begins with the dimensions of a scale drawing. To do a detailed drawing of one part of a project, a team member must know exactly how that part will fit with the others. So teams must communicate and discuss ideas.

When the detailed drawings come together, every team member should examine all the parts. Here's where artistic vision gets tested. Teams must visualize how the project will—and won't—work. If designated team members have not included all the components for their part of the project, the team won't be able to catch potential problems.

## Move to the Computer?

Although most design projects begin with sketches done by hand, designers and engineers often move from pencil to computer for more exact and detailed images. Computer-assisted design (CAD) programs help designers and engineers reproduce images precisely and to scale.

Your computer's word processing program probably has some very basic CAD tools, usually found under the toolbar. (If your drawing toolbar isn't visible, try "View" and "Toolbars" and then select "Drawing.") Although options are limited, this toolbar does allow a designer to replicate shapes exactly. To explore this option, try creating a simple car. First, choose the rectangle and stretch it out (using the program's "handles") until the figure looks like a car. Then try to draw four identical wheels. If you have difficulty making them identical, try an iteration: Do one wheel and then hit "Copy" three times. You can now move the identical wheels to their proper places on the car.

You also can explore perspective with even the simplest drawing programs. A spiral icon on many programs allows designers to rotate an image. Are the four wheels showing up on only one side of the car? Look for "Order" under the "Draw" tools, and change the order of the images on the page.

This exercise illustrates one feature—and one big problem—with a simple drawing program. Each image is actually stored separately by the computer in individual layers on the computer screen. These individual layers take up a lot of computer memory and exhaust disk space or printing capability pretty quickly.

To avoid this problem, designers use a simple CAD program. CAD drawings allow designers to combine images, flip or rotate them, and combine separate shapes into one coherent image. CAD programs also provide drawings done to scale automatically.

In business and industry, CAD programs are linked to computer-aided manufacturing (CAM) tools to create models and actual prototypes. These tools are very useful in the sketching process—helping designers produce, for example, scale models suitable for engineers to evaluate and test.

## Grab Your Pencil

The key to sketching is practice—you'll get better as you do more drawing. Don't be shy. Draw your project again and again, and listen carefully to the comments of your team. If there is a part they don't understand, think about how you can add more detail.

Leonardo da Vinci's sketches have been treasured for centuries because they give us clues on how a great inventor developed his ideas. What clues can you give for the technology of tomorrow?